The valiant Monkey Couch Guardian ready for action.

PIR motion sensor

LED

Resistor, 220Ω

Arduino Uno microcontroller

T0334838

**6. On guard!** When the Monkey Couch Guardian is switched on, any motion within about 20 feet will trigger the PIR sensor. The monkey will start shrieking and clanging his cymbals, scaring away any unwanted intruders, furry or otherwise. You can use the Monkey Couch Guardian to guard a whole room or just your stash of snacks.

And now that you know how to use an Arduino, you can easily adapt this circuit to switch on lights or appliances in your home, yard or workshop, by substituting the appropriate relay.

—*Keith Hammond, MAKE Projects Editor*

To see full build instructions, schematics, breadboard layout and project photos, visit the project page for this build: **radioshackdiy.com/project-gallery/monkey-couch-guardian**

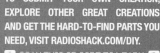

# Make: Volume 32

78

## ON THE COVER

**HAMMER TIME:** Shawn Thorsson poses with one of his Warhammer 40K (Games Workshop) costumes, made from vacuum-formed plastic. Photograph by Cody Pickens. Art direction by Jason Babler.

64

Vol. 32, Oct 2012. MAKE (ISSN 1556-2336) is published quarterly by O'Reilly Media, Inc. in the months of January, April, July, and October. O'Reilly Media is located at 1005 Gravenstein Hwy. North, Sebastopol, CA 95472, (707) 827-7000. SUBSCRIPTIONS: Send all subscription requests to MAKE, P.O. Box 17046, North Hollywood, CA 91615-9588 or subscribe online at makezine.com/offer or via phone at (866) 289-8847 (U.S. and Canada); all other countries call (818) 487-2037. Subscriptions are available for $34.95 for 1 year (4 quarterly issues) in the United States; in Canada: $39.95 USD; all other countries: $49.95 USD. Periodicals Postage Paid at Sebastopol, CA, and at additional mailing offices. POSTMASTER: Send address changes to MAKE, P.O. Box 17046, North Hollywood, CA 91615-9588. Canada Post Publications Mail Agreement Number 41129568. CANADA POSTMASTER: Send address changes to: O'Reilly Media, PO Box 456, Niagara Falls, ON L2E 6V2

WHEE!

# Make: Volume 32

READ ME Always check the URL associated with a project before you get started. There may be important updates or corrections.

**94**

⌃ **SMOKIN':** Prepare delicious meat and fish by building our Nellie Bly Smoker from a 55-gallon barrel.

**147**
**ZIP CAR:**
Make this fast toy car out of laser-cut wood and inline skate wheels.

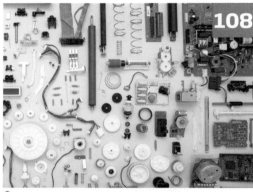

**108**

⌃ **TRASH BECOMES TREASURE:** A computer printer is loaded with reusable components.

# Jens Dyvik explores frontiers of open source, parametric design.

**Jens Dyvik** is an honors graduate of Design Academy Eindhoven in the Netherlands, and winner of the Willy Wortel prize (for Most Innovative Design) among others. Passionate about his research in personal manufacturing and open source design, Jens spends his time design-ing, making, and lecturing at FabLabs throughout the world.

**Jens' Layer Chair is an ongoing open source project.** An interface for the parametric design makes it possible to adapt the chair to different profile curves and material thicknesses. *You can download the free design files at dyvikdesign.com.* Shown in photo above: finished Layer Chairs milled from black MDF.

**Jens recently shared the project at FabLab Sevilla.** Three groups of students customized their own versions of it: one inspired by the classic Eames lounge chair, another, the "perfect" ergonomic chair for architecture students, and finally a version that involved weaving a seat from thin rope.

**Jens on the digital fab revolution:** *"Instant collaboration and implementation are the keys of this revolution. It's about placing digital tools in the hands of free-spirited individuals, who aren't afraid to try and fail and try something else. This lowers the risk surrounding new product development. It's incredibly exciting!"*

| | |
|---|---|
| **WHERE** | Oslo, Norway |
| **BUSINESS** | *dyvikdesign.com* |
| **SHOPBOT** | PRS alpha 96 x 48 at FabLabs all around the world |

*"The ShopBot 96 x 48 is my favorite tool in the FabLab, because it enables me to proto-type anything so well, that a prototype becomes Production Item #1."*
Jens Dyvik

Give us a call to discuss your production needs; we'll help you choose the right ShopBot for you.

> "Things which are different in order simply to be different are seldom better, but that which is made to be better is almost always different."
> —Dieter Rams

**Make:**®

FOUNDER & PUBLISHER
**Dale Dougherty**
dale@oreilly.com

EDITORIAL DIRECTOR
**Gareth Branwyn**
gareth@makezine.com

MAKER-IN-CHIEF
**Sherry Huss**
sherry@oreilly.com

**EDITORIAL**

EDITOR-IN-CHIEF
**Mark Frauenfelder**
markf@oreilly.com

PROJECTS EDITOR
**Keith Hammond**
khammond@oreilly.com

SENIOR EDITOR
**Goli Mohammadi**
goli@oreilly.com

TECHNICAL EDITOR
**Sean Michael Ragan**

ASSISTANT EDITOR
**Laura Cochrane**

STAFF EDITOR
**Arwen O'Reilly Griffith**

EDITORS AT LARGE
**Phillip Torrone**
**David Pescovitz**

WEBSITE
WEB PRODUCER
**Jake Spurlock**
jspurlock@oreilly.com

**DESIGN & PHOTOGRAPHY**

CREATIVE DIRECTOR
**Jason Babler**
jbabler@oreilly.com

SENIOR DESIGNER
**Katie Wilson**

SENIOR DESIGNER
**Michael Silva**

ASSOCIATE PHOTO EDITOR
**Gregory Hayes**
ghayes@oreilly.com

VIDEOGRAPHER
**Nat Wilson-Heckathorn**

**MAKER FAIRE**

PRODUCER
**Louise Glasgow**

MARKETING & PR
**Bridgette Vanderlaan**

PROGRAM DIRECTOR
**Sabrina Merlo**

**SALES & ADVERTISING**

SENIOR SALES MANAGER
**Katie Dougherty Kunde**
katie@oreilly.com

SALES MANAGER
**Cecily Benzon**
cbenzon@oreilly.com

SALES MANAGER
**Brigitte Kunde**
brigitte@oreilly.com

CLIENT SERVICES MANAGER
**Sheena Stevens**
sheena@oreilly.com

SALES & MARKETING COORDINATOR
**Gillian BenAry**

**MARKETING**

SENIOR DIRECTOR OF MARKETING
**Vickie Welch**
vwelch@oreilly.com

MARKETING COORDINATOR
**Meg Mason**

**PUBLISHING & PRODUCT DEVELOPMENT**

DIRECTOR, CONTENT SERVICES
**Melissa Morgan**
melissa@oreilly.com

DIRECTOR, RETAIL MARKETING & OPERATIONS
**Heather Harmon Cochran**
heatherh@oreilly.com

BUSINESS MANAGER
**Rob DeMartin**
rdemartin@oreilly.com

OPERATIONS MANAGER
**Rob Bullington**

PRODUCT DEVELOPMENT ENGINEER
**Eric Weinhoffer**

MAKER SHED EVANGELIST
**Michael Castor**

COMMUNITY MANAGER
**John Baichtal**

ADMINISTRATIVE ASSISTANT
**Suzanne Huston**

**PUBLISHED BY**
O'REILLY MEDIA, INC.
**Tim O'Reilly,** CEO
**Laura Baldwin,** President

Copyright © 2012
O'Reilly Media, Inc.
All rights reserved.
Reproduction without
permission is prohibited.
Printed in the USA by
Schumann Printers, Inc.

**Visit us online:**
makezine.com

**Comments may be sent to:**
editor@makezine.com

**CUSTOMER SERVICE**
cs@readerservices.
makezine.com

**Manage your account online, including change of address:**
makezine.com/account
866-289-8847 toll-free
in U.S. and Canada
818-487-2037,
5 a.m.–5 p.m., PST

**Follow us on Twitter:**
@make      @makerfaire
@craft      @makershed
**On Google+:**
google.com/+make
**On Facebook:** makemagazine

CONTRIBUTING EDITORS
William Gurstelle, Mister Jalopy, Brian Jepson,
Charles Platt, Matt Richardson

CONTRIBUTING WRITERS
Tim Anderson, Thomas J. Arey, Scott Bedford, Phil Bowie
Larry Cotton, Kevin Craft, Stuart Deutsch, Limor Fried,
Saul Griffith, Eric Hansen, Laura Kiniry, Bob Knetzger,
Steven Kotler, Ryan P.C. Lawson, Ed Lewis, Steve Lodefink,
Gordon McComb, Forrest M. Mims III, Meara O'Reilly,
Bob Parks, Dieter Rams, Rick Schertle, Donald Simanek,
L. Abraham Smith, Julie Spiegler, Crazy Talk, Ervin Tibbs,
Gever Tulley, Robert M. Zigmund

CONTRIBUTING ARTISTS & PHOTOGRAPHERS
Nick Dragotta, Evan Hughes, Bob Knetzger, Tim Lillis,
Rob Nance, Cody Pickens, Damien Scogin, Julie West

ONLINE CONTRIBUTORS
John Baichtal, Michael Castor, Michael Colombo,
Chris Connors, Collin Cunningham, Adam Flaherty,
Brookelynn Morris, Nick Normal,
John Edgar Park, Sean Michael Ragan

TECHNICAL ADVISORY BOARD
Kipp Bradford, Evil Mad Scientist Laboratories,
Limor Fried, Joe Grand, Saul Griffith, William Gurstelle,
Bunnie Huang, Tom Igoe, Mister Jalopy, Steve Lodefink,
Erica Sadun, Marc de Vinck

INTERNS
Eric Chu (engr.), Craig Couden (edit.), Max Eliaser (engr.),
Isabella Ghirann (engr.), Gunther Kirsch (photo),
Ben Lancaster (engr.), Courtney Lentz (mktg.),
Miranda Mager (sales), Brian Melani (engr.),
Tyler Moskowite (web), Paul Mundell (engr.), Nick Raymond (engr.),
Josie Rushton (engr.), Daniel Spangler (engr.)

# CONTRIBUTORS

**Julie West** (*Fast Toy Wood Car* and *Label-Etch a Glass Bottle* illustrations) spent her childhood taking art classes and volunteering at an art gallery. She studied art and graduated with a BFA in illustration. She started out as a traditional media artist, but after a few years working as a print and web designer, various computer- and design-related elements began to surface in her work, which is often quirky, bizarre, or ironic. Julie focuses on the way people live and define themselves, and the environments they create, revealing the unique moments that make us human. She now spends her days in her studio, quite content that she makes things all day.

As a kid, all **Gordon McComb** (*Keyless Lock Box*) wanted to do was sit in his den with his trusty cocker spaniel by his side, smoke a pipe, and write how-to articles for DIY magazines. Though he's penned over 1,000 articles and some 60 books since the late 70s, he doesn't smoke, has never owned a cocker spaniel, and works out of a very overcrowded workshop that he shares with three computers and an aging oscilloscope. Among Gordon's better-known books is the best-selling *Robot Builder's Bonanza*, now in its fourth edition. When not writing, he likes watching British TV crime shows, and with his wife he has launched a new line of humorous and geek-centric apparel.

**Evan Hughes** (*Soapbox* illustration) is an illustrator specializing in bold, cartoon-tainted, often surreal, flat-color ink drawings with a bias toward old printing processes. He lives in Scranton, Pa., with his wife and three children. He has been drawing competitively ever since his second grade class voted another student's rendition of Bugs Bunny better than his. This triggered a chain of events that led him to the School of Visual Arts in New York City and the belief that he can draw professionally. He tends to take creative inspiration from blue whales, unsafely built tree forts, Hieronymus Bosch, and Lego creations his six-year-old son makes.

**Gunther Kirsch** (photography intern) has learned more in the past six months at MAKE than he learned from his entire scholarly career. Every day his skills are pushed and expanded. His first day, he encountered a man gathering papers from the printer. Not knowing who he was, Gunther introduced himself, asking, "What is it you do here at MAKE?" The man answered with a laugh, "A lot of things." Gunther found out later the man was Dale Dougherty, founder of MAKE. This taught him that you sometimes learn things by making a fool of yourself. He's found that the quicker he drops his ego, the quicker he learns, and thus the more satisfied he becomes.

**Matt Richardson** (*Get Started with BeagleBone* and *Awesome Button*) wears a lot of different hats: video producer, writer, maker of things, technology consultant, and ITP student. It wasn't always so easy; he admits, "Growing up, I had a lot of trouble teaching myself electronics. It wasn't until a few years ago, when I ordered a Getting Started with Arduino Kit from Maker Shed, that the floodgates opened for me." His most recent project, the Descriptive Camera, "uses crowdsourcing to output a text description of what you snap instead of a photo." He loves running and swimming and lives in Brooklyn, N.Y., with his partner, Andrew.

**Max Eliaser** (engineering intern) is a computer programmer. He can't help it; he was born that way, and he's doomed to spend the rest of his life counting cache misses. He's lucky enough to live in Sebastopol, Calif., just 10 minutes away from O'Reilly Media headquarters. For the last 14 months, he has worked as an engineering intern at MAKE Labs, but he will soon be transferring to Oregon State to study computer science full-time. He'll leave behind his hometown, his cat, David, and Screamin' Mimi's, a local ice cream parlor, but he's excited about pursuing his education. In his spare time, he enjoys scuba diving and writing about himself in the third person.

# Our New Look, By Design

**By Mark Frauenfelder**

Think about the products you love: a favorite kitchen appliance, a mobile phone, a chair. Chances are you admire them not only for their functionality but for their appealing design, too.

In the maker world, design often takes a backseat to the giddy thrill of just making something work. But in the last few years we've seen that attitude change. Ready access to laser cutters, 3D printers, and CNC machines (along with easy-to-use software to design and make things) have sparked a new trend toward beautifully designed and manufactured "maker goods."

Inspired by what we've seen at hackerspaces and Maker Faires around the world, we explore "design for makers" in this issue. Our creative director, **Jason Babler**, decided it was time to redesign MAKE to reflect this new renaissance in DIY art and design. I asked him a couple of questions about our new design, and here's what he had to say.

**MF: Why was MAKE due for a redesign?**
**JB:** Before we talk about why we redesigned, we should talk about why we *restructured*.

We realized new readers could be confused by the mini-branded sections scattered throughout. We had multiple formats for projects, and we wanted to bring order to the book and make it more accessible.

We focused on simplicity in our approach, eliminating confusing sections and merging redundant ones. We also dropped our rigid design templates and made the stories more lively and varied. Prominent page labels now ground the reader instantly, especially when flipping through our beefy issues. Section openers are clearer, and we're letting our designers show off their design chops with the opening spreads in the feature section. The magazine will be familiar to current readers, but we're throwing them some surprises.

**MF: Our website was redesigned ... er, restructured, too. What was the thinking behind it?**
**JB:** The old design was a bit Frankenstein-ish and never felt fully developed. I wanted to establish more visual hierarchy. We also streamlined the navigation and categorized the content to be more intuitive.

Community is also important to us, and our new site design reflects that. We encourage the entire maker community to interact with us by submitting ideas for articles, projects and kits at blog.makezine.com/contribute.

**MF: Besides being the creative director, you're also a terrific character sculptor. I love the cool monsters you've sculpted out of clay. Do you have any tips for makers who want to make their projects look and work their best?**
**JB:** I've met a lot of people who are intimidated by the design process, but it's just about identifying the problem and systematically arriving at a solution. It's eerily similar to the very skills you needed to make the project itself.

Do you see a mess of wires in your project, and want to hide them? Now you're thinking about design. Don't like the size, shape, or function of an off-the-shelf product? Now you're researching machining and production and finishing. You're doing what any designer is paid to do: dream, analyze, plan, and make something new.

And the biggest bit of advice? You learn so much from your mistakes; don't work so hard to avoid them. ◪

Mark Frauenfelder is editor-in-chief of MAKE.

# Punk science, loudspeaker breakdown, fun kid projects, and a father-and-son story.

» In Volume 31's "Sound-O-Light Speakers" project (makeprojects.com/project/s/2464), two diagrams on page 82 describe an acoustic suspension (sealed) alignment and a bass reflex (vented) alignment. Unfortunately, the description of the bass reflex design is based on a popular but incorrect theory — that the back wave of the driver is made available via a port or opening to augment the sound from the front of the driver.

In reality, the port is a Helmholtz resonator whose resonance is defined by its length and cross-sectional area. The port is typically tuned to a frequency that's lower than that of the driver — it's not a manifestation of the back wave of the driver.

The opening at the bottom of the Sound-O-Light speaker is not a port since the entire chamber would be the resonating column. It more closely resembles a transmission line speaker.

Vented designs are by their nature less forgiving and more difficult to work with. For a hobby-type speaker, a sealed design is far easier to build and get right.

I plugged the HiVi B3N driver's parameters into a Thiele-Small calculator (you can find these online). In a sealed alignment, for a speaker box volume of 0.1ft³, the length of the 3" pipe is a manageable 24.49" and the max bump in low-end response is 0.98dB. This will add a little boominess to the sound and some false impact to the bass, and should be OK.

A vented alignment is not so easy. The optimum calculated volume is 0.59ft³, for a max low-end bump of 1.63dB. The vent (port) will be 1"×1.5" (or 1.5"×4.85" but that's hard to fit in this design), tuned at 36.9Hz. Unfortunately, our 3" pipe would have to be over 12' long! If we reduce the volume to 0.1ft³, the low-end bump rises unacceptably high, and the port tuning has to be moved up, so we lose the extended low end, and the port's output can't match the bump; the result is a really bad response curve.

In summary, a small speaker made from this driver works best in a sealed box. If we allow for significantly more volume (a bigger box or a 3" pipe that's 12' long), the vented alignment can be made to work and yield a much better low end than the sealed box. It's a tradeoff.

Or, one could ignore the details and just build the thing and enjoy it. :)

—*Louis Lung, lungster.com,*
*Westborough, Mass.*

**Projects Editor Keith Hammond replies:**
Louis, thanks for your excellent analysis. We thought we were clever to offer a vented option but it's clearly trickier than we knew. Consider yourself an honorary member of the MAKE Technical Advisory Board, Loudspeaker Division.

》 Just had to tell you how much I loved Gareth Branwyn's Welcome in MAKE Volume 31 ("Three Test Tubes and the Truth"). You perfectly captured the intersection between DIY and punk. I feel like I have a new understanding of my relationship with both. One more reason to raise your kids punk!

—R. Mark Adams, Westport, Conn.

》 We built the MAKE Compressed Air Rockets Kit (makeprojects.com/project/m/2235) from the MAKE School's Out Summer Fun Guide (makershed.com/schoolsout) and it took about 2½ hours to complete the launcher. We came back to make the rockets another day. It was a total blast, kid safe, and the rockets shoot amazingly far. We lost a few to the trees, so we angled the launcher to shoot down our street — and easily shot 100 yards. Great project!

—Jack Lamb, Santa Clara, Calif.

》 We had a ball making the "Marshmallow Shooters" project (makeprojects.com/project/m/2000) from the MAKE School's Out Summer Fun Guide on the weekend — and they were so simple. Now we have 10 neighborhood kids wanting to build the same thing. We might have to run a workshop for them all. Cheers, MAKE.

—Earl Pilkington, Mandurah, Western Australia

》 In January 2011, I read Len Cullum's Japanese-style "Workhorses" project (makeprojects.com/project/w/572) and thought they were beautiful. At the time, I had taken on only simple projects that required no more than a miter saw and a cordless drill. Your sawhorses would introduce me to many new skills, so I decided to make them. They would also allow me to spend some quality time with my father, a very experienced woodworker who was declining in health rapidly due to pulmonary fibrosis.

After 17 months, I just finished my project! However, a few things happened along the way. Most importantly, my father received a lung transplant in May 2011 and has had a full recovery. Secondly, your beautiful sawhorses became the "ends" of a pretty good-looking table (see below). I probably spent 400 hours working on the table and about 30 hours talking to my father on the phone about it.

I turn 40 this year and this table represents a lot for me. My kids hated it because it took so much of my time, but now they love it. My 8-year-old son thinks this project had to be on my "bucket list"!

Thank you so much, Len, for contributing that article to MAKE.

—Walt Stewart, Decatur, Ala.

**Author Len Cullum replies:** Maybe this happens to MAKE contributors all the time, but I am blown away that someone would build my project and then take the time to share their story with me. I can't find adequate words to describe what it means to me. All I can say is thanks to you, Walt, and to everyone at MAKE for giving me the opportunity. I am humbled.

## MAKE AMENDS

In Volume 31's "Sound-O-Light Speakers," the vented build option is not a true bass reflex design (see reader Louis Lung's letter, opposite). Also, author Bill Gurstelle's Night Lighter 36 spud gun project (makeprojects.com/project/s/5) is from MAKE Volume 03, not 04.

In Volume 31's "Mendocino Motor" project, the correct URL for downloading the 3D part files is makeprojects.com/v/31.

A caption in Volume 31's "Drive-by Science" incorrectly identified Dr. Akira Sugiyama's affiliation; he's with Tokyo University.

# The {Unspoken} Rules of Open Source Hardware

**By Phillip Torrone,** Open Source Enthusiast

I truly believe open source hardware is here to stay. It has established itself as a great community, a great effort, and for many, a great business. I spend most of my days working on open source hardware in some way, and I want to talk about some of the unspoken rules we all seem to follow.

Why? Because the core group of people who've been doing what we've collectively called "open source hardware" know each other — we're friends, we overlap and compete in some ways, but we all work toward a common goal: to share our work to make the world a better place and to stand on each other's shoulders and not each other's toes. Some folks will agree strongly and others will disagree. That's great — it's time we start this conversation.

### We pay each other royalties, even though we don't need to.

As odd as it sounds, we pay each other. I can be really specific: I introduced Mitch Altman, inventor of the TV-B-Gone, to Limor Fried, owner of Adafruit Industries. I wanted to convince him to work with her on a kit version. That was almost five years ago and it's worked out great. Mitch travels the world doing workshops while MAKE, Adafruit, and many others sell his kits, and he gets paid a royalty.

Behind the scenes, open source hardware designers pay a royalty to each other when they make and design together. Do they need to? Technically no, but we all do. Actions speak the loudest for this unspoken rule.

### We credit each other, a lot.

What does the open source maker usually want? Just to be credited properly. This usually isn't an issue since the community members look out for each other, but there are examples where it's just unclear who made what. It's usually not malice, just forgetfulness.

There are a lot of giant companies taking open source ideas and making them commercial products (that's always going to happen), but the open source hardware community is a community. We credit each other. When we get general ideas, we usually say things like "This was inspired by such-and-such." Giant companies don't or can't do this, but the open source hardware world can.

### Naming: It's better to be unique.

In general, we try to avoid naming our projects in a confusing way. Trademarks are one of the few ways we can "intellectually protect" hardware (schematics are not copyrightable), so we focus on branding things and building products that people know come from a specific company or person.

There was a period where many people and companies made Arduino-like boards and stuck "-uino" at the end of the name or even just called their product Arduino, but I see that ending soon. The Boarduino, for example, was OK'd by the Arduino team. This was before a million "-uinos" came out later.

More and more makers are creating new and unique Arduino compatibles and calling them something completely non-"uino."

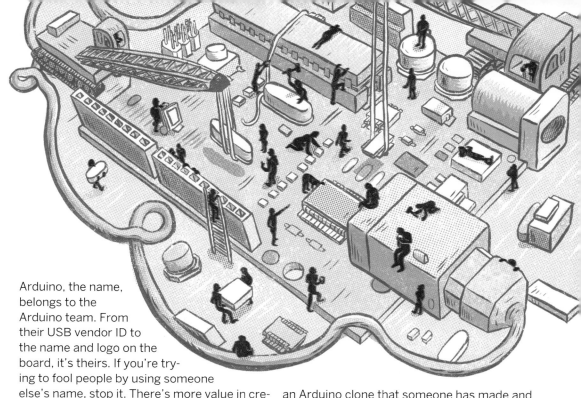

Evan Hughes

Arduino, the name, belongs to the Arduino team. From their USB vendor ID to the name and logo on the board, it's theirs. If you're trying to fool people by using someone else's name, stop it. There's more value in creating your own name for your own products.

## We actually do open source hardware.

This is an easy one. If you're calling it open source hardware, release the files — schematic, source, BOM, and code — all under an open license. Don't hide them. Don't say you need to sign an NDA and attempt to obfuscate. If you're trying to be tricky, just don't do open source hardware.

Open source hardware isn't a marketing term — it means something specific. We're doing open source hardware because we want to, not because we want to trick people. The only issue that usually comes up for us at Adafruit is time, as we manage hundreds of projects concurrently, so not every file is updated instantly, but I'm going to try my best to make sure they're all up. I'm moving everything to GitHub to make this easier on me (and everyone).

## Basing your project/product off open source? Open source it.

Let's say you make something based on an Arduino, which is under an open license — you need to do the same. Once in a while I'll see an Arduino clone that someone has made and put under a non-commercial license. When I ask why, it's usually something like, "Well, I don't want to be cloned — like the Arduino is all the time." Sometimes the maker changes the license to an open one after the project makes the rounds. If you do an Arduino shield it should be open source hardware too.

## Code and designs: Add value.

It's not valuable for the community to fork code and just change a name or something small and call it your own. You need to add more value. Many open source hardware companies have really expensive teams making and sharing open source code and hardware. Just changing a couple of things so you can ship your own thing is really frowned upon. It happens, but it's pretty rare. It's one thing to copy and improve — it's another to just copy and sell.

I'm a big fan of copy, improve, and republish, but it's rarely done because it's hard work. When people fork just so they can change one comment or make it sound like they're the original authors, it's a support burden for the original makers too. It can be a mess. For open source hardware to work, we all need to support the original authors when we can. We want to avoid people or companies building their products off the open source software/hardware communities and then closing them off. Sharing needs to go both ways, always.

### Cloning ain't cool.

If your goal is just to make Arduino clones and not add code or hardware improvements, please go do something else instead. I see a few companies that make straight-up clones, give them confusing names, and think it's socially acceptable. It's not. The beginners get confused as to what's a real Arduino with the quality, service, and support, and most of the time the clones are crappy.

I have a box of "Arduino killers" from all over the world. They're not adding value in any way; they're just examples of people being selfish. I get a dozen emails a week from parents or kids who bought a fake Arduino and are upset it doesn't work and that the eBay seller or fly-by-night store won't help them. Most of all, if any reasonable person gets cloned enough, she might just stop doing open source due to the support burden.

### Support your customers.

If you're doing open source hardware because you want to make an Arduino clone thinking you can just pass the hard work of customer support over to the community, that's not fair to anyone. Spend the time and resources to create tutorials and forums, and support your customers. I'm using Arduino again as an example since I see customers purchasing cloned Arduinos but expecting support from the Arduino support team because it says Arduino. Open source is a way to make things better, not to just outsource support to someone else. Join in, support your customers, and they'll reward you.

### Build your business around open source hardware.

If you're going to require that someone does open source for your newly venture-funded online open source hardware social network or whatever, you gotta do some open source yourself. If you're celebrating open source and attempting to make money around it, you gotta put value back in too.

For example, if part of your product design is requiring customers to have all their files under an open source hardware license, you need to do that too and open up your own stuff. Otherwise, what's the point?

Obviously there's marketing value in the word "open," and for small startups we've seen that many want to take advantage of that. Want your new company to be part of the open ecosystem? It's worth something, so you need to do the same. I'm not saying you need to give it *all* away, but you need to do something to show you value open source enough to do it yourself.

### Respect the designer's wishes.

Sometimes the maker of an open source hardware project might have a request if you're going to clone their hardware; for example, "Hey, don't use this to kill puppies, OK?" Now, while open source really doesn't stop anyone from making a puppy grinder from your open source CNC, it's totally fair for the designer to ask you not to do that.

A few times, I've seen an open source hardware project get hijacked a little, and the author was concerned about its direction. We can email each other and talk when needed. It's a strength that we're a community with members who can talk to each other. It's also helpful for the designer to include a bit of text in a Readme for the license or on a project page that lists some ideal uses. Of course it won't always be followed, but at least there's some framework and intentions spelled out. We are humans who get emotional about our works; it's not a weakness. This, too, is a strength. ◪

➕ Share your thoughts and read the full version of this column at makezine.com/go/osh_rules.

Phillip Torrone is an editor-at-large of MAKE and creative director at Adafruit Industries, an open source hardware and electronic kit company based in New York City.

# That's right. 2 for 1.

Share your passion for making with everyone on your list with MAKE's 2 for 1 holiday offer

# Protecting Your Ideas

**By Ryan P. C. Lawson, Esq.**
Small Business Advocate

As a maker, your ideas and what you make are important to you. You might want to start selling your idea and make money. Or maybe you want to give your idea away, as long as people also contribute their improvements for free. Either way, you may want to protect it with intellectual property protections.

Unfortunately, you can't protect something if it's just a thought. If you can creatively express your idea, you can protect it with a *copyright*. If you can turn it into an invention, you can protect it with a *patent*. If you keep your processes secret and sell the result, you can protect it with a *trade secret*.

Copyrights protect an original expression in a tangible medium. This could be a song, an image, a story, a set of instructions, or even computer code. Your idea itself isn't copyrightable, but the expression of it is. If you give me a written expression, I can use that idea as long as I don't substantially copy the expression. For example, if you have an idea to build a faster computer chip and give me the directions, I can build the chip without violating your copyright.

Patents can be used to protect any new and useful process, machine, or composition of matter. This definition doesn't include "idea." To patent a process, it must produce a tangible result. Having a patent means that you can prevent anyone else from using or making your invention, even if they're only using the invention in their own home. Registered with the U.S. Patent and Trademark Office, patents can cover new designs for 14 years or have some new utility to them and last 20 years. Obtaining them can be a fairly complex process, and you may want the help of a specialized patent lawyer if you're serious about obtaining one.

A trade secret is something that your business knows that isn't known to the public. Your idea is only protected from someone wrongfully disclosing it. There's no protection from someone independently discovering it. The benefit is that you can keep your idea secret, but there are no protections from someone discovering it on their own. If I bribe a Coca-Cola executive to give me the formula for Coke, Coca-Cola could sue me for violating their trade secret rights. If I'm a good chemist, though, and experiment until I work out the formula, then I wouldn't have violated Coca-Cola's trade secret.

If you want to give your ideas away using an open source license, you'll need to make sure they're protected first. Other than trade secrets, intellectual property can't be used without some sort of a license or legal provision that allows someone else to use it. You can give away a license if you want. As part of giving the license away, you can put conditions on it. For example, I might say you can use this license as long as you don't sue me. You could say that people can use a license to your intellectual property as long as they give away any improvements for free. Unless you have something to license, though, you can't force anyone to agree to the share and share-alike terms.

You can also sell a license. When you buy a book, piece of music, or software program, what you're actually buying is a license to use the copyrights associated with that work.

If you want to sell your ideas or create a community of sharing around them, you'll want to make sure you properly protect them first. What kind of protections you use will depend on the form your idea takes. If you're in doubt as to how to protect your idea, it may be worth your time to consult a lawyer who specializes in intellectual property. ◪

Ryan Lawson is a Michigan lawyer. His practice focuses on technology licensing and advising small businesses.

Our favorite events from around the world.
Compiled by William Gurstelle

# Maker's Calendar

## International Snow Sculpture Championships
### Jan. 22-26, Breckenridge, CO

Perhaps the world's greatest snow sculptures are carved out of giant, dense blocks of Rocky Mountain snow 10 or more feet on each side and weighing 20 tons. The competition begins at 11 a.m. with a shotgun start and continues for four days. makezine.com/go/snowsculpture

NOV DEC JAN

## Festival della Scienza
### Oct. 25-Nov. 4, Genoa, Italy

It's the 10th anniversary of one of Italy's premiere science-related festivals. Explore interactive science and art exhibitions, workshops, performances, and more. festivalscienza.it

## TeslaCon
### Nov. 30-Dec. 2, Madison, Wis.

Steampunk lovers gather at TeslaCon to revel in all things neo-Victorian science fiction. There are enough wonderful contraptions, costumes, and accessories to satisfy all. teslacon.com

## International Space Station Expedition 34 Launch
### Dec. 5, Baikonur Cosmodrome, Kazakhstan

An international crew of American, Canadian, and Russian astronauts rocket off to the ISS, weather permitting, aboard an old-school Soyuz spacecraft. Once there, the crew will test how the human body adapts to long periods in space. makezine.com/go/expedition34

## Santiago Mini Maker Faire
### Dec. 15-16, Santiago, Chile

STGO Makerspace hosts South America's first-ever Maker Faire! STGO will showcase work from their members (40 and growing) as well as the greater Santiago community. makerfairesantiago.com

## Nothing Much Happens Day
### Dec. 21, Worldwide

Despite the end of the Mayan calendar and claims of the mysterious planet Nibiru crashing into Earth, nothing out of the ordinary happens today. makezine.com/go/dec21

## Louisiana Fur and Wildlife Festival
### Jan. 13-14, Cameron, La.

Those who make the trek to the remote hamlet of Cameron for "one of the oldest and coldest festivals in Louisiana" can increase their backwoods knowledge by learning and improving their oyster shucking, trap setting, muskrat skinning, and other Cajun-revered skills. lafurandwildlifefestival.com

✱ IMPORTANT: Times, dates, locations, and events are subject to change. Verify all information before making plans to attend. Visit makezine.com/events to find classes, fairs, exhibitions, and more. Log in to add your events, or email them to events@makezine.com. Attended a great event? Talk about it at forums.makezine.com.

# Transparent Exploration

**Mika Aoki**'s glass-blown sculptures seem otherworldly, until you look more closely. Then you realize how very tied to this world they are: inspired by mold spores, viruses, plants, life, and death, the clear glass simultaneously makes abstract concepts concrete and vanishes before our very eyes. Born in Hokkaido, Japan, Aoki now lives in London, where she is learning English and studying at the Royal College of Art. (Don't mistake her for a novice; she's shown her work for years in Japan after studying art and glass blowing in Tokyo.)

By ***Arwen O'Reilly Griffith***
Photos by ***Yoshisato Komaki***
**sing-g.net**

Is she an artist or a scientist? Her resumé proves her chops as an artist, but she says, "My inspirations come from observations and conversations with scientists." While creating new pieces, Aoki often visits labs to spark ideas, and dreams of collaborating with scientists on her next body of work. Her interest in things that can't always be seen with the naked eye lends itself perfectly to glass. "Unless light shines on it, we cannot confirm [its] existence," she writes. "Thus I encountered one material which can exist as the membrane of something invisible."

In an installation piece called *Her Songs Are Floating*, an old car sits in darkness. Glass arches out of the car and within it, looking like transparent roots shooting into the interior. "I try to make works that could make contact with physical and mental senses," Aoki says, and one can't help but think of the battle between humans and the natural world, life and its end. Other works show sinuous glass sculptures suspended in vitrines, exploding from test tubes, and growing out of bottles. Sperm searches for ovum, virus for host, spore for sustenance. Surely the glass is alive? Or at least singing? "I'm interested in the phenomenon of life," she says simply.

# The Stuffing of Dreams

By *Gregory Hayes* Drawing by *Sophie*

**childsown.com**

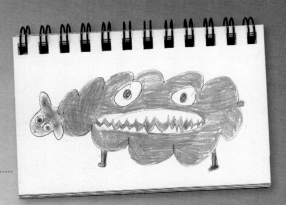

Asked to send a comfort toy with their child to kindergarten, many parents would have hit the toy store for a plushie. **Wendy Tsao** decided to make one.

"I was planning to make a favorite animal, but then I saw my son's self-portrait," says Tsao. "It was something he drew all the time, and always the same way," she says. "After I made it, he recognized it immediately and was very pleased and appreciative. That's when I knew I was on to something."

This revelation kicked off Tsao's small business, Child's Own Studio, in Vancouver, B.C. Now she brings children's artwork to huggable life as toys full-time.

Self-taught, she learned from borrowed library books and looking at lots of stuffed toys. "I made lots of softies and made lots of mistakes," she recalls. Her unabashed DIY approach hasn't stopped the charming toys from becoming a runaway success, though.

While she is surprised at how many adults want toys made from their own drawings, Tsao prefers working with kids. "[Adults' drawings] intimidate me with all their slick and demanding details. Children's drawings allow me more space for creativity and inspiration, which makes it a real collaboration between artists.

"Happily, I think I have finally found my niche and I can't imagine doing anything else."

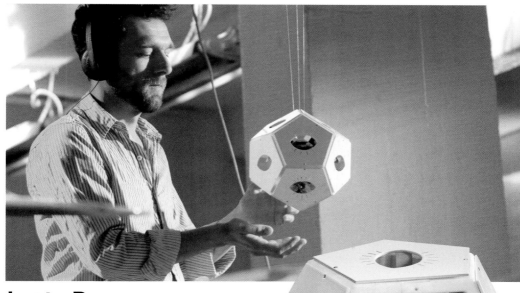

# Let's Dance

By *Craig Couden*

**vimeo.com/28651568**

Step away from the turntables and the sea of knobs and switches and use the rhythm of your body to drop some beats with the Dodecaudion music controller.

Developed by **Jakub Kozniewski**, **Piotrek Barszczewski**, and **Krzysztof Cybulski** of the Polish art collective panGenerator, Dodecaudion is a gesture-based, spatial interface that allows performers to create music by moving hands or other body parts around the 12-sided structure. Each face uses an analog, infrared distance sensor connected to a custom Arduino shield, which collects positional data and sends it wirelessly to a computer via Bluetooth. A bridge program converts incoming data to Open Sound Control, which can be input into your favorite synthesizer. OSC has the ability to create sound and visuals simultaneously, one of the trio's goals for the project.

Kozniewski's TEDxWarsaw talk showed the device's potential to move beyond typical electronic music. During the talk, performer Maddie Bovska interacted with Dodecaudion using her hands, arms, shoulders, stomach, back, legs, and toes in a way more akin to modern dance than a Friday night house party.

"Most of the time, gestures are a byproduct of performing live music. We asked what could we do to change that," Kozniewski explained to the TEDx audience.

The Warsaw-based artists began developing Dodecaudion in 2010 using breadboard prototypes, but really got going a year later when they partnered with Hedoco, a company that helps Polish innovators develop open source products. Code for Dodecaudion is available at Hedoco's website, and users are encouraged to share and remix their creations.

"We strongly believe that open sourcing a product is the way to make it evolve and develop." Kozniewski declares.

# Shredding the Chopper

By *Laura Kiniry*
**brother-and-sister.com/
guitars/guitars.html**

When it comes to constructing outrageous instruments, San Francisco musician **Michael Gaughan** has truly broken the mold. The 32-year-old builds his own guitars and plays them on tour with his two-person metal/punk/rock band, Brother and Sister. His guitars aren't your run-of-the-mill six-strings: one has an entire ant farm in its body, and his Helicaster — built using a gas-powered R/C helicopter that can hold and lift up to 5 pounds of electric guitar parts safety — can actually fly. "The entire piece is hooked up to a wireless radio unit," says Gaughan, "so it can produce feedback and noise when it's both flying *and* played."

But Gaughan also has an appetite for destruction. "I thought it would be really funny to be playing a guitar that you could eat," he says. After a few ill-fated attempts (including a Rice Krispies-and-chocolate guitar that held together for only one song), he perfected hard candy guitars, the ultimate sugar rush.

To make them, Gaughan fashioned a mold out of aluminum roofing, duct tape, and tinfoil. He then mixed corn syrup, water, food coloring, and 40 pounds of sugar and boiled it on the stovetop at 350°F (an earlier try at 250° left the floor covered in candy), poured it in the mold, and let it harden before adding guitar parts.

Gaughan made eight one-time-use guitars in red, yellow, green, and blue. "We would make them and then rush to the show, play time, let people lick them, smash them, and then eat them," he says. "They were treated like objects for a ceremonial ritual of rock 'n' roll."

Sure, the guitars are a bit unsanitary, admits Gaughan. "[But] they sound sweet."

Rik Sferra (guitars), Tony DeRose (Viper)

# Fast & Furious Fuselage

By *Goli Mohammadi* **the-viper.org**

Amid the sea of projects at Maker Faire Bay Area 2012, one shining standout was crafted by a team of five young makers, all under the age of 18. Welcome to the Viper, a full-motion flight simulator built into the fuselage of a Piper PA-28 plane, complete with 360° rotation on both the pitch and roll axes and a fully immersive flying environment inside. Not your typical after-school project.

Team Viper is **John Boyer** (17), **Joseph DeRose** (13), **Sam DeRose** (17), **Sam Frank** (17), and **Alex Jacobson** (17), all members of the Young Makers club. Inspired by a simulator at the National Air and Space Museum, they set out to build a better version based on *Battlestar Galactica's* Viper spaceship.

Mission accomplished. Once the rider is harnessed in the Recaro racing seat with a full helmet, the plane door is put in place. Inside the cockpit, three 22" high-def screens display the game *FlightGear*, which you play as you fly. The armrests hold the joystick and thruster, while the custom instrument panel, dozens of buttons and LEDs, and sound system complete the full immersion experience. For control the team used five Arduinos, two iPhones, and one iPad, all networked together. As Sam D. says, "The only senses we don't control are taste and smell – that's for Maker Faire 2013."

*Follow us @make* **25**

# A Way with Wood

By *Arwen O'Reilly Griffith* **christophfinkel.com**

**Christoph Finkel** comes from a long line of woodworkers (four generations long!). Growing up in the German Alps exposed him both to a traditional skill set and to "beautiful, big, old trees." His father still works as a wood turner, making *rodels* (German sleds) the traditional way. "I often got the wood that was left and tried to make nice things out of it," Finkel recalls.

He took a slight departure from his forefathers by studying sculpture at the Academy of Fine Arts in Nuremberg, but says, "Even though I learned a lot of new things at the

Academy, I got the basic understanding and fascination for wood from my dad."

His love of wood is expressed in masterfully carved vessels that are profoundly indebted to the wood they're carved from. "It's a sympathetic and warm material compared to iron or plastic," he says. Elegant bowls, sliced so that they hardly seem able to hold together, curve around knots and burls. Handsome vessels with precise geometry suddenly tilt to accommodate natural shifts in the wood. Most importantly, each vessel highlights both the craftsmanship and the tree.

# Tall Drink of Water

By *Laura Cochrane* **makezine.com/go/tropismwell**

» Instead of cricking your neck to drink from a water fountain, what if the fountain cricked its neck for you? This thought occurred to designers **Richard Harvey** and **Keivor John**, after seeing a call to rejuvenate public drinking fountains. » Friends since age 10, Harvey, 27, and John, 28, run Poietic Studio, a London design company. (*Poietic:* productive, formative. From the Greek *poiētēs:* maker, poet.) Their combined expertise includes audiology, interaction design, and repairing classic porches. » The pair developed their idea into Tropism Well, an interactive sculpture that senses when someone is near and bows to pour water into a glass. The first prototype employed a linear actuator to bend the neck, but the movement was too robotic. Then they realized the weight of the water could be used to achieve a more natural bowing motion. » The final iteration is almost 10 feet tall, with a base of stacked wooden discs, reminiscent of a spine, and a stainless steel neck with a glass carafe on the end. An Arduino Mega and ultrasonic sensor work to detect the user, triggering water to pump up into the carafe. As the weight at the top increases, the Well gracefully bends its neck, pouring water if the sensor installed on the carafe detects a waiting receptacle. » "We see it like a generous mother goose," Harvey explains. "It's using the weight to power the movement rather than a motor; that gives it the feel of something more natural and something you can have empathy with. We think this is why people say 'thank you' to it."

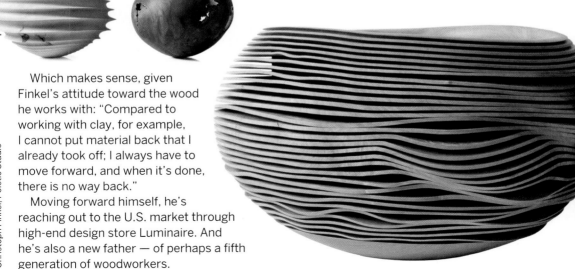

Which makes sense, given Finkel's attitude toward the wood he works with: "Compared to working with clay, for example, I cannot put material back that I already took off; I always have to move forward, and when it's done, there is no way back."

Moving forward himself, he's reaching out to the U.S. market through high-end design store Luminaire. And he's also a new father — of perhaps a fifth generation of woodworkers.

## EXCLUSIVE INTERVIEW WITH SHAWN THORSSON

# MAD PR

Shawn Thorsson makes toys for people who reject the notion of growing up. If you bump into one of his 8-foot-tall Space Marines, be sure to apologize.

Interview by *Goli Mohammadi*
Photography by *Gregory Hayes*

## What was your first elaborate costume build? What inspired you to make it?

A set of Stormtrooper armor from *Star Wars*, which I made in 2002. I'd been stationed in Japan with the Navy for two years and was really excited to be back in the States for a proper Halloween again. After watching Kevin Rubio's fan film *Troops*, I wanted nothing more than my own suit of white armor. The problem was that I'm a little short for a Stormtrooper and all I came across were suits designed to fit someone around 6 feet tall.

Since these [in *Troops*] were fan-made suits, I reasoned that as long as someone else could figure out how to make it, there's no reason I couldn't figure it out. Then I happened across studiocreations.com, which featured a very detailed how-to for making your own vacuum-formed Stormtrooper armor just like they did in 1976 for the original film. Six months and two nearly catastrophic garage fires later, I was out on the town winning Halloween costume contests in every bar and club I went to.

## How do you choose a character to build?

I don't have any particular criteria. If something strikes me as cool-looking, I'll find out more about it. I look around to see if anyone else has already made it, and whether or not I think I can do it better. I've been trying to pick projects that show off my skills and let me work with different materials and textures.

The biggest thing I'm aiming for lately is diversity. After I built a few copies of my Stormtrooper armor, I put together a pretty rushed Boba Fett costume and then decided

I needed to do something very different before I became just another *Star Wars* costume guy. That's what drove me to make the *Predator* costume and the Spartan outfit from *300*. Then, in 2010, I finished the *Red vs. Blue* costumes and a whole arsenal of different weapons to go with them. It was a great project and got a lot of attention, but I rushed to do another, something different, so I wouldn't become just another *Halo* costume guy.

## How did you learn your fabrication skills?

When I was a kid, my grandfather often told me that you can learn anything anyone anywhere knows from a book. I spent a lot of time in the library back then. Nowadays, the same thing can be said about the internet. Just about anything you could want to know can be found somewhere online, but it can also be easy to get discouraged. The trick is to find inspiration and ideas without allowing yourself to be overwhelmed by the volume of other folks out there who have already done what you're thinking of trying.

Aside from that, I've learned from trial and error more than anything else. This usually means that I know all sorts of technical things without having any of the proper terminology to describe them. I also spend a lot of time looking at materials and thinking of what they could be used for, above and beyond the manufacturer's recommendations.

Believe it or not, my fabrication skills actually aren't all that extensive. I'm a high school shop teacher's worst nightmare.

## How did you develop the 16-step painting process for aging your pieces?

Painting and finish work is usually my favorite part of a build. It's not always quite that elaborate, but if I have luxury time at that stage,

I really like to have fun with it. I suppose it's the same thing that comes up at the end of a recipe where they write "season to taste."

By now, my finish work employs everything from mustard to hairspray to rock salt, and nothing is ever painted with just one color. The end result is an amalgamation of additive and subtractive processes using spray cans, paintbrushes, steel wool, stencils, toothpicks, and the airbrush along the way.

## What are your three most indispensable tools?

I have a spacious shop, a small fortune in equipment, and no shortage of tools. Still, the three things I get the most use out of are my head, my hands, and my friends. But if I've already got those, there's a handful of other things I couldn't do without. First is probably my Dremel rotary tool. I burn one up every six or eight months. I also have a Craftsman CarveWright, essentially a very simple CNC machine. Aside from those, I do a surprisingly large amount of work with nothing more than a sharp knife.

## Your costumes are showstoppers. Describe some of your favorite reactions.

When you walk out in public looking like you just stepped off the screen from a popular movie or video game it's often like becoming an instant celebrity.

Back when *Halo: Reach* was released, I went to the midnight launch and showed up at the local GameStop in a dead-on, screen-accurate version of the Master Chief's armor from *Halo 3*. As I approached the line of folks waiting for the release, one kid, probably about 12 years old, walked up and asked, "Can I have a hug?" That threw me a little. I'm covered in full battle armor, nothing friendly about me at all, and this kid wanted a hug? So I say, "Sure." Next thing I knew, there's a line

Cody Pickens

■ **More from Thorsson at** makezine.com/go/thorsson.

of fans waiting to hug me. That was a bit weird.

There was also an episode in the Stormtrooper armor that was pretty fun. That Halloween I was in one particular nightclub, walking across the dance floor. When I reached a gap in the crowd, the DJ cut the music, hit me with a spotlight, and called out "TK-421, why aren't you at your post?"

## What inspires you to share build instructions?

I suppose I'm just a team player. I learned everything I needed for my first big build from someone who just decided to share a few tutorials years ago. Hopefully I'll inspire someone else somewhere along the way.

## What's next?

I'd like to say I have a plan, but mostly I've just got ideas. There's been a lot of discussion with my friends about doing one or more group builds. I'm also trying to think of a way to make something bigger than my 8-foot-tall Space Marine costumes. I'd also like to dabble a bit in animatronics so I can add more life to some of my projects. I've just started accept-

ing professional commissions to build costumes, props, and set pieces for low-budget films and promotional use. So really, there's no way to know what will come up next.

## You're also currently in the Navy, correct?

Correct. I'm a lieutenant commander in the Navy Reserve. So I attend drills one weekend a month and go out for a couple of weeks of annual training every year. Most of the time it's uneventful, but once in a while something comes up, like last year when I got to help out with the relief effort in the wake of the tsunami in northern Japan, or a few years ago when I was sent to Afghanistan to work with the provincial reconstruction teams rebuilding key infrastructure like roads, running water, and electricity.

## What do other sailors think of your creations?

Most of the time folks seem impressed. More often than not, one of them will tell me something like, "I showed your webpage to my kids and they think you're the coolest guy in the world." That's always great to hear. ◪

**VIDEO GAME DEVELOPER HI-REZ STUDIOS COMMISSIONED SHAWN THORSSON** to re-create the Spinfusor weapon from their online shooter *Tribes: Ascend* for the PAX Prime show in Seattle. Pictured below is one of the reference shots they sent him, and here's the final weapon in the hands of Blood Eagle Pathfinder, one of Thorsson's full costume builds. See how he went from 3D file to rough CNC-cut MDF pieces to completion. Full build details at **makezine.com/go/spinfusor**.

# MAKING THE
# SPINFUSOR
# FROM TRIBES:
# ASCEND

Shawn Thorsson and Matt Herman (opposite)

**CHARACTER**

Isaac Clarke

**VIDEO GAME**

Dead Space 2

Cast resin parts built onto a custom vest and a heavily modified pair of coveralls. Armed with a Plasma Cutter from Epic Weapons. **Total build time:** Four months. **Completed:** 2011

**CHARACTER**

Imperial Space Marine

**TABLETOP GAME**

Warhammer 40,000

**All Space Marines:** Vacuum-formed styrene and ABS plastic. Details and helmets made of urethane resin. **Total build time:** Five months. **Completed:** 2011

**CHARACTER**

Spartan: Sister

**WEB SERIES**

Red vs. Blue (Halo)

**All Spartans:** Peparuka models used as sculpting armatures. Molds taken and parts cast in urethane resin. Face shields custom-formed in orange-tinted acrylic, mirrored to match character. **Total build time:** Three years. **Completed:** 2010

**CHARACTER**

Spartan: Dexter Grif

**WEB SERIES**

Red vs. Blue (Halo)

Also known as Minor Junior Private Negative First Class Dexter Grif, he is a known slacker and loudmouth, famous for saying, "Why does everyone think I'm yellow? Seriously! Didn't anyone have a box of crayons when they were a kid?"

**CHARACTER**

## Sister Repentia

**TABLETOP GAME**

## Warhammer 40,000

Artist's rendition of Sister Repentia character. Mask parts and chainsword vacuum-formed in ABS plastic with cast resin details on boots and chainsword. **Total build time:** A few hours. **Completed:** 2012

**CHARACTER**

## Spartan: Doc

**WEB SERIES**

## Red vs. Blue (Halo)

Also known as Medical Officer Super Private First Class Frank DuFresne, his purple armor is a combo of red and blue, since he helps both teams. Seen here carrying a plasma pistol by Jasman Toys.

**CHARACTER**

## Imperial Guards

**TABLETOP GAME**

## Warhammer 40,000

Helmets rotocast in urethane resin, and remaining armor is vacuum-formed ABS plastic. Lasgun and Laspistol by fellow prop builder Matsucorp. Note squad number 707, Thorsson's area code. **Total build time:** Three weeks. **Completed:** 2012

**CHARACTER**

## Imperial Space Marine

**TABLETOP GAME**

## Warhammer 40,000

The popular 8-foot-tall Space Marines are genetically modified superhuman soldiers, the elite warriors of the Imperium of Man. Shown here holding the signature "bolter" firearm.

**CHARACTER**

## Spartan: Sarge

**WEB SERIES**

### Red vs. Blue (Halo)

Pictured with the MA5C Assault Rifle, since Sarge's trademark shotgun was left in the rafters at Thorsson's workshop. His preferred nickname is S-Dog. "You just got Sarge'd!"

**CHARACTER**

## Imperial Space Marine

**TABLETOP GAME**

### Warhammer 40,000

Genetically modified to have super-human strength, speed, endurance, and resistance to injury, enhanced senses, reduced need for sleep, and poisonous saliva secretions that can degrade metal over time. Watch your back.

**CHARACTER**

## Predator

**MOVIE**

### Predator

Muscle suit made of upholstery foam and latex, details of duct tape, aluminum bar stock, latex, cast resin, polymer clay, leather. **Total build time:** Two months. **Completed:** 2005

**CHARACTER**

## Spartan: Caboose

**WEB SERIES**

### Red vs. Blue (Halo)

Armed with the BR55HB Battle Rifle, Private Michael J. Caboose of the Blue Team is a central character, known for being eccentric, dimwitted, and divorced from reality. "Buttons! Oh man, I love buttons!"

**CHARACTER**
## UNSC Marines

**VIDEO GAME**
## Halo 3

Background soldiers. Can be seen in fan film, *Halo: Helljumper,* available at helljumper.com. **Total build time:** One month. **Completed:** 2011

**CHARACTER**
## UNSC Marines

**VIDEO GAME**
## Halo 3

"When I die, please bury me deep! Place an MA5 down by my feet! Don't cry for me, don't shed a tear! Just pack my box with PT gear!" —UNSC Marching Song

**CHARACTER**
## Blood Eagle Pathfinder

**VIDEO GAME**
## Tribes: Ascend

Mostly rotocast urethane resin with a small fortune in LEDs built in. **Total build time:** Three months. **Completed:** 2012

**CHARACTER**
## Kali

**VIDEO GAME**
## Smite

Costume parts and accessories made of latex, urethane foam, and urethane resin. **Total build time:** Three months. **Completed:** 2012

# Workshop
## SHAWN THORSSON

Down a dusty road in the eggs-and-butter town of Petaluma, Calif., sits a big, handmade pink Victorian complete with chicken coop, storage containers, vintage cars, and Thorsson's badass workshop, where fantasy comes to life.

**1.** Homemade frakkin' hot vacu-forming oven **2.** Drill press **3.** Band saw **4.** Vacuforming table **5.** MDF for CNC machine **6.** Assault rifle mold **7.** Space Marine boots **8.** *Halo* Spartan boot molds **9.** Bucket of silicone **10.** Chainsword **11.** Rare patch of open floor **12.** Combat garden gnome molds **13.** Freshly painted helmets **14.** *Halo* replica rifles **15.** Another drill press **16.** The back burner **17.** Fridge

# The Titan of Toy Invention

Written and illustrated by **Bob Knetzger**

**Marvin Glass was the man behind Mouse Trap, Operation, Lite-Brite, Rock 'Em Sock 'Em Robots, and dozens of other iconic toys and games.**

His innovative products changed the world, yet he was never satisfied and always working on the next thing. He was a flamboyant showman and delighted in cleverly presenting his latest top-secret ideas. He earned the fierce loyalty of his employees, although he could be exceedingly difficult, even abusive. He was a genius when it came to design, but he didn't invent his most famous products himself. His personal life was complicated and moody, but his creations brought joy to millions. And after his untimely death, his most well-known products live on: plastic barf, wind-up chattering teeth, and the games Mouse Trap and Operation. Obviously, we're not talking about Steve Jobs — this is the story of Marvin Glass, the titan of toy invention.

Baby Boomers may not have known his name, but Glass' amazing toys and games and their TV-promoted jingles were unforgettable: "His block is knocked off?!" (Rock 'Em Sock 'Em Robots); "Open the door for your — *ahhhh!* — Mystery Date!"; "Don't touch the side! *Bzzzzt!*" (Operation). Some Glass toys are distant memories, but many remain popular and still sell well today.

Born to German immigrants in 1914, Glass grew up near Chicago. During an unhappy childhood he created his own toys from cardboard and wood: a toy dog, swords and shields, and a climb-inside toy tank. Foreshadowing a lifelong pattern, his creations made his friends happy, yet he remained lonely.

He designed animated store window displays after college and sold an invention for a toy theater to a manufacturer for $500. When he learned the company made much more from his

Mouse Trap (Ideal Toys, 1963)

# Hits (and Flops)

Glass revolutionized the game business with 3D game/toy hybrids, most famously Mouse Trap, still made today.

**Crazy Clock:**
A sequel to Mouse Trap (Ideal Toys, 1964)

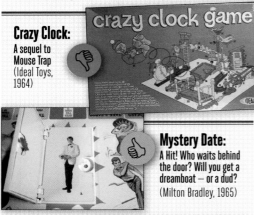

**Mystery Date:**
A Hit! Who waits behind the door? Will you get a dreamboat — or a dud? (Milton Bradley, 1965)

**Fish Bait:**
Another try for a Mouse Trap-style game. (Ideal Toys, 1965)

**Tiger Island:**
A reaction time game with a spinning tiger and marble-tossing castaways — stranded in the toy aisle! (Ideal Toys, 1966)

**Perils of Pauline:**
Sculpted and painted 3D game components can't rescue a boring game (Marx Toys, 1964)

idea ($30,000!), Glass realized what he must do in the future: license his designs and earn a royalty on each one sold.

An inventor named Eddy Goldfarb brought his invention, Yakity-Yak Teeth, to Glass. They struck a deal: Goldfarb did the inventing and Glass did the selling. Glass sold the design to novelty maker H. Fishlove and Company of Chicago, and the comical wind-up choppers became a hit in 1950. Glass promoted other Goldfarb creations — and himself in the process. Goldfarb told me, "Marvin was a great, *great* salesman. He should have been in the movie business!" (Goldfarb eventually left Glass to create hundreds of toys on his own, like Mattel's Vac-U-Form and Schaper's Stomper mini motorized cars. Goldfarb was a true pioneer of the invention part of the toy business.)

Glass created a lavish, top-secret toy studio in Chicago worthy of Willy Wonka. Instead of Oompa-Loompas he hired talented designers, inventors, sculptors, and model makers. Lathes and mills were painted in bright colors. Chagall paintings and Remington sculptures filled the waiting rooms at Marvin Glass and Associates (MGA). Every new toy or game idea was kept absolutely hush-hush. Soundproofed windows and triple-keyed locks, all monitored by closed circuit cameras, prevented corporate espionage. Prototypes were locked up in a safe every night.

Was Glass' paranoia justified? Probably not, but the mystique of such dramatic flourishes only added to the perception by visiting toy company executives: these ideas must really be something! According to writer and toy inventor Richard Levy, a big part of salesmanship is: "It's not what you have — it's what they *think* you have."

And what ideas they were! MGA's new toy concepts had clever features that often created all new categories of toys and games.

Children's board games had been a sleepy product category with such staid offerings as Candyland and Chutes and Ladders. "Bored" game was right. MGA changed all that with their innovative skill and action games. Rock 'Em Sock 'Em Robots took its inspiration from an arcade boxing game but was miniaturized for tabletop action and added a telegenic gimmick: landing a solid push-button punch on the robot's jaw made his head pop up with a revving "buzzzzzzzz!" sound.

MGA's game and toy hybrids ("gamoys" in the toy biz lingo) transformed the flat game board into three dimensions. Mouse Trap was inspired by Rube Goldberg comics, masterfully rendered in styrene gears, ramps, and chutes, along with springs, rubber bands, and marbles. The gameplay wasn't much more than ritualized assembling of the toy contraption. No matter. The real fun was activating the chain reaction of mechanical gizmos that ultimately dropped a cage on a mouse token, ending the game.

A huge hit in 1963, Mouse Trap

Marvin Glass with Ideal's Robot Commando, 1961.

what you think.

MGA's clever design featured a wired remote control with a Bowden cable inside. When you twisted the knob to move the pointer from "Forward" to "Shoot," a stiff cable pushed or pulled to mechanically shift gears inside the robot. At the same time, the vibrations of your voice sympathetically jiggled a metal contact, which connected the battery power and energized the motors. You could say "Fire!" but if the knob pointed to "Turn Left" the robot turned left. Dramatic, successful, and artfully deceptive: Robot Commando was not unlike Marvin Glass himself.

Toy manufacturers like Ideal, Hasbro, and Louis Marx lined up to get a peek at MGA's latest inventions. The string of hit toys seemed endless: take-apart,

---

*Glass created a lavish, top-secret toy studio in Chicago worthy of Willy Wonka. Instead of Oompa-Loompas he hired talented designers, inventors, sculptors, and model makers.*

---

is still sold today, virtually unchanged. MGA attempted to clone its success with two sequels, Crazy Clock and Fish Bait, but the timing was wrong and kids didn't bite. That's the fickle toy biz!

Robot Commando was a great example of a Glass toy: the exciting 1961 TV commercial shows a giant, motorized robot that listens and obeys. Say "Forward!" into the mouthpiece control and the robot goes forward. Say "Fire!" and he shoots missiles from his domed head or flings balls from his spinning arms. Wow! What boy wouldn't want a robot that responds to verbal commands? But it's not

put-together robot Mr. Machine (1960), slap-action card game Hands Down (1964), toss-the-hot-potato game Time Bomb (1964), glowing, colorful peg-picture maker Lite-Brite (1967), freewheeling Evel Knievel Stunt Cycle (1973). You couldn't turn on a TV, open a Sears Christmas catalog, or go into a kid's bedroom without seeing an MGA design. ◪

Bob Knetzger (neotoybob@yahoo.com) is an inventor/designer with 30 years experience making fun stuff. He's created educational software, video and board games, and all kinds of toys from high tech electronics down to "free inside!" cereal box premiums.

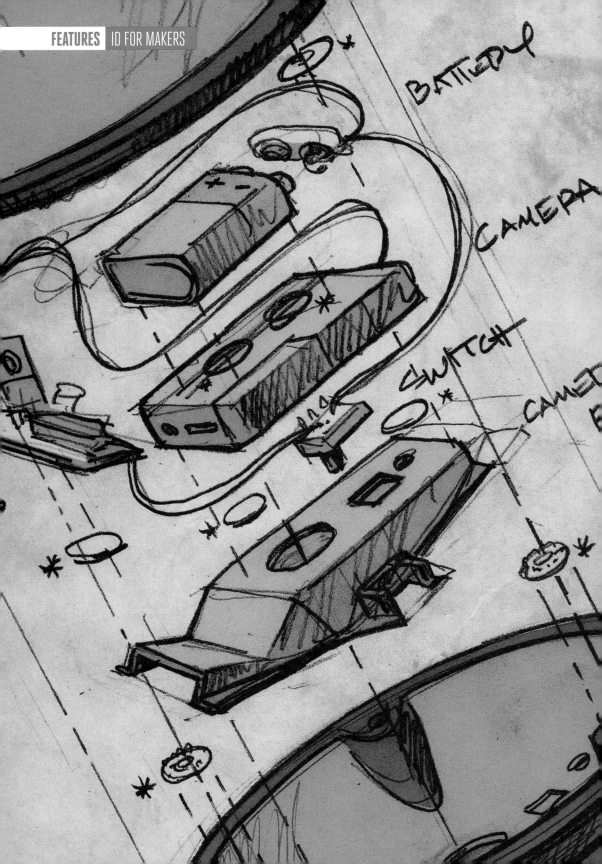

BATTERY

CAMERA

SWITCH

CAMERA

# Industrial

Written and illustrated by **Bob Knetzger**

## *Design*

## for Makers

**OK, so you've made your project. Whether it's an original concept of your own or something from a MAKE how-to article – good for you! Maybe you've learned a new skill or tried a new tool or process: that's the fun and reward of making.**

But consider your end result. How well does it really work? How does it look? What would make it better? Maybe you're ready to take your creation to the next level: manufacture and sell. How would a professional maker approach it? That's a job for an industrial designer!

Industrial design (ID) is the science and art of creating commercial products, experiences, and environments. The skills and techniques the designer uses include ideation sketching, drawing/ rendering, drafting, sculpting, and model making. An industrial designer also must know about materials, manufacturing processes, electronics, computer programming, engineering, printing, and graphics.

The industrial designer is the champion of a new product and sees the concept through all the stages: presenting new ideas to management, selling them to marketing, proving the design to engineering, and shepherding the design through the legal and patent processes. The designer needs to consider product safety, cost and ease of manufacture, and package and logo design for print, web, or TV advertisements. In many ways, an industrial designer is a "professional maker."

## HISTORY OF ID

Since humans began making things, from flint-edged tools to flintlock rifles, we've been designing. But in the 20th century, modern manufacturing and mass marketing required a new combination of talents. In the fast-paced, competitive marketplace, only the best-looking, best-working, and most affordable products (and the companies that made them) survived. Manufacturers sought out individuals who had the vision and skills to make the many decisions required in producing a product.

One of the first and most famous industrial designers was Raymond Loewy. An engineer by training, his natural artistic skill and dramatic French flair earned him important commissions. In 1929, Loewy, a true maker and DIYer, hand-sculpted clay in his NYC apartment living room to create a sleek, simplified design for Gestetner's mimeograph machine. His streamlined designs made things cleaner and easier to use, as well as better selling. Loewy and his firm went on to create or redesign many iconic forms of 20th-century America: the Coca-Cola bottle, the Pennsylvania Railroad S1 locomotive, the Studebaker Avanti, Air Force One, and many others. He famously said, "The most beautiful curve is a rising sales graph."

In the second half of the 20th century, husband-and-wife team Charles and Ray Eames expanded the field of industrial design to include not just products, but also exhibits (IBM at the World's Fair) and educational films (*Powers of*

(Top) Raymond Loewy's design for the Studebaker Avanti, 1961. (Bottom) Eames bent plywood litter for U.S. Navy, 1941.

*Ten*). Together, they created handsome chairs and furnishings for homes and offices. Also very hands-on, the Eames assembled their own electrically heated and bicycle pump-pressurized molding machine, nicknamed the "Kazam!" They used it in their Los Angeles apartment to bend flat pieces of plywood into three-dimensional shapes, creating splints and litters for the U.S. Navy. The classic Eames chair is the direct descendent of their early experiments. Their application of both science and art produced tasteful and elegant designs that connected with consumers emotionally and continue to sell and inspire designers 50 years later.

Today, industrial designers like Jonathan Ive (Apple iMac, iPad) or Yves Behar (XO laptop) have access to advanced technologies like Cintiq tablets, SolidWorks, and rapid prototyping fabrication, but the basic approach to

The original Balloon Cam.

*Quick notations spread over the page as ideas flow forth, like visual stepping stones.*

designing a new product is the same.

As an exercise, let's take a sample MAKE project and see how the creative process used by an industrial designer might improve it.

## THE ID CREATIVE PROCESS

The Helium Balloon Imaging "Satellite" camera from MAKE Volume 24 is Jim Newell's clever hack made by adding a timer circuit to the shutter button of an inexpensive camera, then sent aloft by tethered helium balloons to take aerial photographs. It's made of repurposed items: a CD serves as a platform to hold the camera, a prescription bottle contains the circuit and battery, and it's all held together by twists of wire. It looks

as if it were thrown together from items scavenged from the trash bin: MacGyver-worthy, but not a very elegant design.

It's also not very easy to use: you have to remove the camera (it's fastened to the CD with double-sided tape), turn it on, replace the camera, connect the circuit board, attach the battery that starts the timer, reconnect the cap (and don't let the balloons get away while you're doing all this!), let out the string to fly the balloon, taking pictures as it goes up. It's begging for a redesign.

For the purpose of this article, let's assume there's a client who wants to manufacture and sell the Balloon Cam as a DIY kit and is working with an industrial designer to help refine the design.

Sam Murphy

The design process includes these steps:
1. Define scope
2. Ideate
3. Review
4. Develop prototype
5. Test and revise

## 1. DEFINE SCOPE

The first steps are the most important and set the direction the product will take. The client and designer agree on all the constraints and problems that need to be solved by the final design. Some concerns are for the manufacturer, some for the consumer. For this project, the revised design should be:

» **Inexpensive to manufacture**, meaning low tooling costs. No sense in dreaming up an injection-molded design if the molds will cost too much or take too much time to create.
» **Easy to assemble.** We'll assume the person buying this kit would have some basic skills and tools.
» **Low-cost in its components.** The price of the kit can't be too high, and we'll assume the customer will provide his own hacked camera and circuit.
» **Cool-looking and inspire purchase.** It should be a "wow!" and not a "meh." The design of a product alone can make the sale.
» **Light, strong, waterproof, and easy to fly and take pictures.**

A real project for an actual manufacturer would have many more concerns like product safety, patent infringement, green manufacturing, end-of-use disposal, etc. This list is enough for our exercise.

## 2. IDEATE

Designers use *ideate* to describe the active process they use to generate ideas. To realize and explore directions, designers draw many rough sketches. Quick notations spread over the page as ideas flow forth, like visual stepping stones. These aren't meant to be pretty

pictures or even drawings that anyone else will see: they're used to express and explore ideas. Think Leonardo da Vinci's codices (he even wrote his notes backward so no one else could read them!).

The point of the ideation phase is to create as many related ideas as possible: more ideas will lead you to better ideas. Have a playful sense of humor with the ideas. Don't settle on one idea yet — keep 'em coming!

Here are some of the ideation sketches for the Balloon Cam. Note these design threads:

» **Look for a simple solution:** some kind of shell in a shape that holds and protects the camera and circuit board.
» **Unify the elements:** use the strings to both support and align the components.
» **Simplify features to streamline use:** give access to the switches to turn the

(Clockwise from the left) Preliminary ideation sketches: keep the ideas flowing! Three lobed "ears" provide routing for the rigging. Top shell can be slipped up to access components inside. Eliminate the lobed ears, and add recesses for string harness clearance. Rounder, domed shape. Balloons tethered to docking station with a water bottle as a weight and a hands-free crank.

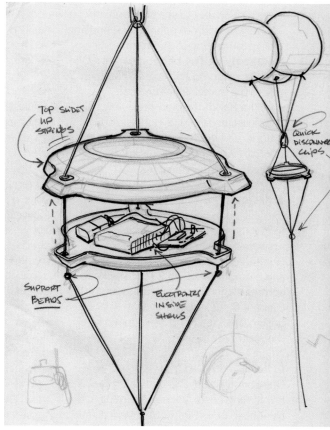

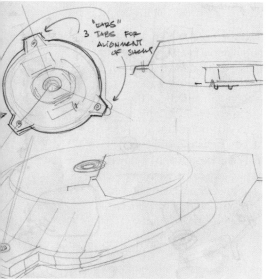

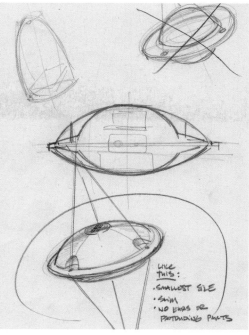

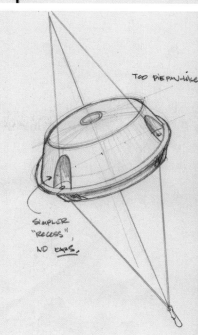

camera on and to start the auto shutter circuit without opening the housing or reconnecting the camera.

» **Make flying the balloon easier:** add a weighted docking station to hold the balloons before flight.

» **Add a hand-cranked reel.** The designer makes notes, corrections, and improvements in sketch form, expressing, evaluating, and re-expressing as he goes.

## 3. Review

The designer selects the best combination of ideas and refines them into a proposed design. This requires more finished drawings, with plans, elevations, and a rendering of what it looks like to communicate the detailed design to the client. If the client approves the paper design, then it's on to developing the prototype.

## 4. Develop prototype

One constraint of this design project is low tooling costs. Vacuum forming is the perfect solution: light but strong plastic shells can be vacuum formed using low-cost patterns. Vacuum forming is also easily doable for the DIYer (*see MAKE Volume 11, page 106, "Kitchen Floor Vacuum Former"*).

The mold makes parts that can be used for both the top and bottom shells. Leaving the flanged rim on one part makes the top shell that fits over a smaller, trimmed bottom shell. It can even be molded in colored plastic so that no painting is required.

Here, drawing is used as a tool to solve a 3-dimensional problem: where to locate the camera, printed circuit board, and 9-volt battery inside the saucer shape. Side elevation and top orthographic views show the location of each component and the minimum size and shape of the basic shell. Further refinements simplify the design and remove the 3 protruding lobes. Instead, recessed features guide the support lines and at the same time provide mating horizontal surfaces to hold the shells together.

The next step is to make the prototype. The domed part of the vacuum-form mold is made from urethane foam, easily turned on a lathe using hand

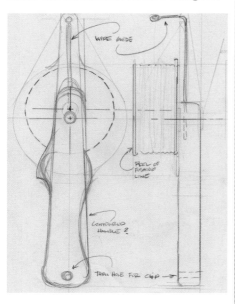

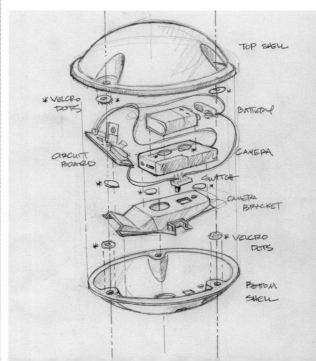

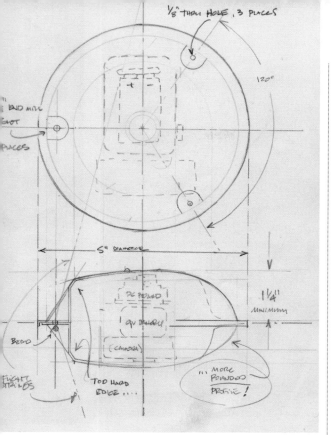

*(sketch annotations)*

⅛" THRU HOLE, 3 PLACES

120°

5" DIAMETER

1¼" MINIMUM

PC BOARD

9V BATTERY

CAMERA

"...MORE ROUNDED PROFILE!"

TOP HARD EDGE.....

BEZEL

FLIGHT STRAKES

END MILL SLOT PLACES

A

B

C

D

(Opposite) Reel details. Exploded view shows assembly and the added vacuum-formed bracket to support the camera and circuit board. (Above) Top and side views drawn in actual size. The curved section side view is used to make a template for turning on the lathe.

tools. First cut a round blank of foam and mount it to the lathe (**Figure A**). Use a cross-section view from the elevation drawing to make a same-size cardboard template gauge for checking the progress of turning the exact shape (**Figure B**). File the shape and check frequently with the gauge (**Figure C**). If you don't have a lathe, you can file the foam by hand, freeform.

Make 3 equally spaced rounded slots with a mill: first, mount the turned foam in a dividing head and mill the slot with a ⅝" end mill (**Figure D**). Index the head by 120°, cut the second slot, then index to 240° and cut again. If your mill has an adjustable head, add 2° of draft for easier unmolding. If you don't have a mill, you can lay out the location of 3 equally spaced holes on the foam with a compass and drill three ⅝" holes. Then carefully cut the slot with a handsaw.

To make the lipped flange feature, cut a second oversized disc from fiberboard and mount it to the foam, **Figure E**. Drill ⅟₁₆" holes all around the edges to provide airflow for snug vacuum forming.

Vacuum-form the shells from 0.030" styrene sheet and trim to check the fit (**Figure F**). The geometry of the curved dome shape, the flanged rim, and the 3 recesses all act together to create a rigid and strong part even when made from thin plastic.

Building preliminary prototypes like this often requires making small revisions. To hold the camera level in the curved shell, add a small bracket made from bits of styrene, solvent-bonded in place. In our theoretical exercise, this bracket would be a third part, vacuum-formed alongside and at the same time as the shells for very little additional cost. See the exploded view drawing on page 50.

To keep costs down, the buyer of the kit would do all the finishing work and assembly:

» Trim the 2 shells and bracket, and solvent-bond the bracket to the bottom shell.

» Drill a ¾" hole in the center for the camera lens (**Figure G**).

» Drill a ¼" hole for access to the camera's mode button, and cut a square hole to view the camera's LCD (Figure G).

» Cut another square hole for the timer circuit's new on/off mini slide (Figure G).

» Drill three ⅛" holes in the flats of both shells for the support strings to run through.

» Fasten the camera and circuit inside the bottom shell with velcro tabs. It's important to be able to remove the camera easily to upload the photos to a PC and change the camera's battery.

E

F

G

H

(Far left) Easy-to-start camera timer.
(Left) Reel clips to the water-filled bottle, and anchors the balloons as the saucer hangs chest-high off the ground, making it easy to check the batteries inside.
(Right) Balloon Cam Success!

» Punch holes in 3 pairs of velcro dots and fasten them around the 3 string holes.

» Make the rigging: tie three 36" lengths of 12lb test fishing line to a split ring. Tie a bead 24" from the ring on each line. Thread each line through a hole in the assembled shell, and then tie the ends together to another ring. Be careful so that the shell hangs level when supported by the 3 beads — that way the camera will face straight down for taking pictures in flight (**Figure H**).

For the working prototype the reel handle is cut from wood and a crank is fashioned from styrene sheet and rod stock. A bent wire form acts as a line guide. A water-filled milk bottle serves as a weight that clips to the handle. Mounting the reel of fishing line off-center looks odd, but works great: when you take your hand off the crank, the line tension automatically stops the reel from spooling out and the balloon stays where it is.

## 5. TEST AND REVISE

The redesigned version of the Balloon Cam is much easier to use for taking aerial pictures: the docking station provides hands-free setup, the camera and circuit are much easier to turn on, and the shells pop open to allow access for changing batteries and download-ing pictures. It even looks cooler with its cute, retro-UFO look as it hovers and flies around.

But how to manufacture the parts for the reel handle in quantities to make affordable kits? Hmm, that's another job for the industrial designer — back to the drawing board! ◪

Bob Knetzger (neotoybob@yahoo.com) is an inventor/designer with 30 years experience making fun stuff. He's created educa-tional software, video and board games, and all kinds of toys from high-tech electronics to "free inside!" cereal box premiums.

By **Max Eliaser** with **Dan Spangler,**
**Eric Chu,** and **Brian Melani**

# Extreme *Makeovers*

## HOW *MAKE* OVERHAULS PROJECTS ...

## Flying Wing R/C Plane

The Towel airplane (MAKE Volume 30) is minimal: a broad wing made of blue foam, with exposed electronics on a Coroplast deck. Dan wanted to make his really pop while adding the durability to withstand Maker Faire demos for years to come. He hid the wiring between the electronics deck and the wing, used Coroplast for the whole wing to further stabilize the craft, and then extended the deck all the way to the nose. This overhaul compromises the reusablility of the deck and makes the nose just a tad more dangerous. But with his final coat of paint, he really likes the new sleek, modern look.

## Dog Ball Launcher

The original Fetch-O-Matic (Volume 31) was housed in a drab plywood box with an unassuming hopper. Dan decided to add an injection of whimsy and a few colorful details to the design: protruding lips on top and bottom, rounded corners, a spiral arrow cutout and MAKE badge, a redesigned hopper, and extra vertical reinforcement. These details add strength and beauty; it may still be a box, but the last thing you'd call it is drab.

**More resources we use in MAKE Labs:**
» Working with plastics: tapplastics.com/product_info/videos
» Woodworking tricks: wood magazine.com/woodworking-tips
» Create detailed fiberglass shells: makezine.com/go/fshells

# Brilliant!

**Making laser-cut acrylic enclosures.**

*By Eric Chu and Dan Spangler*

Acrylic looks great but it's prone to cracking when worked by tools. Solution? Laser-cut acrylic cases. Just send a PDF or other vector image of your pattern to a laser cutter. Here are some tricks we like.

**Finger joints, aka box joints** Great-looking, simple, and strong. CAD up your own, or try boxmaker.rahulbotics.com.

For a sliding finger joint (like for T-slot constructions), use a single line to form the shared edges of a joint. Our laser has a 0.004" kerf (the total width of material the laser removes) so it gives the two adjacent edges a 0.002" cut.

**T-slots** These hold a machine screw and nut for fastening joints. When drawing a T-slot, use the nominal dimensions of your hardware for a sliding fit. Add extra space at the top (like a lowercase t) so you can tighten the screw a little more.

Slot thickness and tooth height should be the same as your material's thickness; measure with calipers. For the width of slots/teeth, I use twice the thickness.

**Living hinges** Amazingly, you can cut a pattern into rigid acrylic to make flexible hinges, aka "sninges" (named for Rotterdam's Snijlab). Learn how to make them at makeprojects.com/project/g/1683.

**Cutting acrylic with hand or power tools** OK, sometimes you must. Learn how at makezine.com/go/cutplastic.

## ... TO MAKE THEM SHINE

# The Electronic Nag

When you walk by the Notification Alert Generator (Volume 30), it reminds you to do your chores. Eric took design inspiration from the elegant lines and restrained color of an old-fashioned refrigerator to give it this modern yet retro look. The parts are laser-cut from white acrylic and screwed together using T-slots (see above), with a speaker grille that spells out NAG. It seems right at home hanging in the kitchen.

# Geiger Counter

This gadget (Volume 29) began as a bare circuit board, which carried the risk of electric shock as well as damage to the circuit. Dan experimented with several styles before choosing this finger-jointed, transparent green box, with grilles to allow the entry of radiation, mounted controls, and handles that let you point the unit at a sample. The result protects the circuit board and evokes a sort of mad-atomic-scientist aesthetic.

Max Eliaser, Dan Spangler, Eric Chu, and Brian Melani are engineering interns in the MAKE Labs in Sebastopol, Calif.

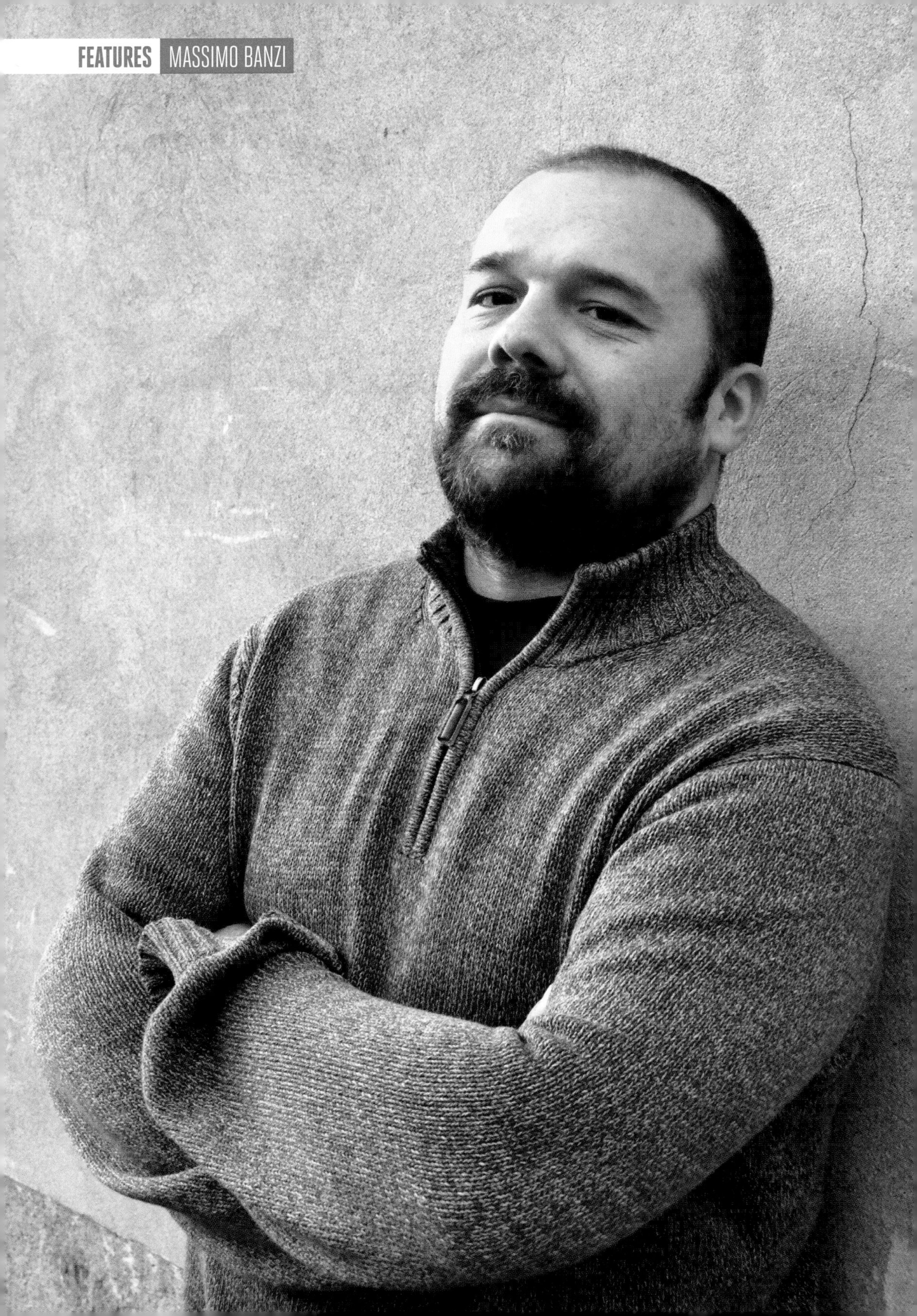

# Good DESIGN Gets Out of the Way

By **Dale Dougherty**

## ARDUINO'S MASSIMO BANZI ON INTERACTION DESIGN

**Arduino is Italian. It was co-created by Massimo Banzi, at Interaction Ivrea, a design school in northern Italy, as a tool for designers to create interactive experiences.**

In March, I spent several days with Massimo, first at a makers conference in Rome and then in and around Turin, where the Arduino is made. We started at the FabLab in Turin, which had previously been part of a large Fiat factory, now vacant. Then we drove out to a countryside area where the early personal computer maker, Olivetti, once thrived.

After the computer maker closed its doors, descendants of Olivetti set up shop to manufacture the Arduino (the microcontroller that has captivated the maker world), which explains why there's a small industry in the Piedmont region with all the machinery needed to make printed circuit boards (PCBs).

On the drive to the factory, I sat in the back of a Fiat and asked Massimo about interaction design, the Arduino, and design for makers.

**Dale Dougherty: What is good design? What can makers learn to make them better designers?**

**Massimo Banzi:** In design there are different fields, and like art, there are different movements. If you look at the design of Apple products, that design descends directly from the Bauhaus and their minimalistic, clean shapes that emphasized rationalism. You see it later in the work of Dieter Rams, a German designer, and Rams' influence on Jonathan Ive (Apple's senior V.P. of industrial design) and Steve Jobs.

Good design is about using the minimal amount of stuff that you need. Also, if something is visually simple, it encourages people to use it.

**DD: What about the intuitive sense of design?**

**MB:** Design can be used to make an object desirable. You can make an object so that others are attracted to it. You might say it's beautiful, that it's striking a chord in you, that you want to have this thing.

David Cuartielles

I tend to think that people who come from an engineering background value adding multiple options, as many options as possible, making the object customizable. In my opinion, design is about finding a way to say "no," deciding which things you're not going to put in the product.

We get a lot of criticism at Arduino because we refuse to apply some of the modifications that users submit to us. As a result, people get disgruntled and they move to other open source projects. We try to explain to them that we're trying to keep the system clean and consistent. We don't want to confuse people.

When I was trying to learn Perl years ago, I saw that you could do the same thing in five different ways. This was completely illogical to me. I said, "Why?" Then I came to understand that in the world of geeky programmers, each one of them has their own preferred way of doing things, and the language had to cater to them.

In design, I think if you have to cater to everyone, you become useless. Nobody will connect to your product. If you design something around yourself or some person, then you'll find some people who will connect to that, because the product you design has a personality.

> *"Good design is about using the minimal amount of stuff that you need."*

**DD: What is interaction design?**

**MB:** I have a fairly light definition. Interaction design is the design of any interactive experience. It can be the interface of an object, say a device with three buttons. That interface can be very bad, an unsatisfying experience, or those same three buttons can create a really nice experience. Everything is an experience, an interaction between you and something, and that experience can be designed. That, for me, is the general definition.

In my case, interaction design tends to be about technology. A lot of the experiences you have today are enabled by objects that contain electronics and sensors. Technology enables the communication between you and the device, or you and a service.

The interaction designer must know design but must also understand technology enough to know what kind of experience you can create with a certain tool.

It's also about understanding business models. You have to understand how the business model is tied in to the experience because it might define what you can create.

**The design path of Arduino, an open source microcontroller for the cost of a pizza.** (Left) Arduino Prototype 0: still called "Programma 2005" as an evolution of "Programma 2003". (Middle) First useable prototype. Still called "Wiring Lite", used as a low cost module for Wiring users. David Cuartielles joined during this stage (the flying resistor is his first contribution to the design) from this point on the project becomes Arduino. (Right) Arduino Extreme v2: Second production version of the Arduino USB board. This is has been properly engineered by Gianluca Martino. Serial number from 501 to (more or less) 2000.

**DD: Is this kind of design more about functionality than aesthetics?**

**MB:** In the world of design, you have different points of view. For some, it's about the visual appearance. For others, it's about functionality. It can be about solving problems. For some people, before you design something, you have to have a debate about the context in which this device will be used, how it will impact people's lives.

One of the classic points of view is that form follows function. Others say that form follows fiction. First you create the story, and then the shape and direction come out of the story, the narrative that you create.

Sometimes technology is described simply in terms of what it can do, where perhaps interaction design focuses on what a human being can do or wants to do with it as a user. Something that's designed purely from a technical point of view doesn't always work — the engineer wants to solve a problem but doesn't go the extra mile to figure out how a person will use it.

**DD: How do you teach interaction design?**

**MB:** It's all about iterations. You start with sketches and prototypes. Then you have different fidelities. You might start with cardboard, as an example of a low-fidelity prototype. When you stuff some electronics in there, the prototype starts to behave like the thing you want to design and so this becomes a high-fidelity prototype.

These iterations allow you to try the prototype with people. You want to figure out how people behave when they're using the thing you've designed. Gillian Crampton Smith, the former director of Ivrea, used to call this the "crafting of the interaction." So you're crafting the interaction in the way an old-style designer would craft the shape of the object.

It has been said that in interaction design, you're not asked to design a vase for flowers. Instead you ask yourself, "How do I arrange flowers around the house, or how do I enjoy flowers and plants in my living space?" During

## Bauhaus Design

Bauhaus (1919–1933) was founded in Germany as a new kind of art institute by Walter Gropius, who wrote the following:

"What the Bauhaus preached in practice was the common citizenship of all forms of creative work, and their logical interdependence on one another in the modern world. Our guiding principle was that design is neither an intellectual nor a material affair, but simply an integral part of the stuff of life, necessary for everyone in a civilized society. Our ambition was to rouse the creative artist from his other-worldliness and to reintegrate him into the workaday world of realities and, at the same time, to broaden and humanize the rigid, almost exclusively material mind of the businessman."

the design process, you might not come up with a vase. You might find yourself designing something completely different. It's good to take a step back and look at things differently.

Where I teach, every class is hands-on. It's always about making a project. Since I believe that interaction design is about trying things with people, the more you want to make your product perfect, the more you need to be able to play with the product. This means that the shorter each iteration, the more experiments you can do. The tools that we use are chosen because of their ability to shorten the loop.

Arduino was successful in interaction design because you could iterate very rapidly around hardware and software until you got something that was working. You could also put it together with other parts.

**DD: What was the path that led you from Ivrea to Arduino?**

**MB:** In the beginning, I started teaching my students the same way I was taught, just copying what my teachers had done. After a couple of lectures, I realized it wasn't going to work. I started to explain what electrons were, Ohm's law, they didn't understand. Then I realized that's not how I learned. The way I learned was through experimenting. When something didn't work, I would go back and try to understand why. So that theory became useful to me, and it matched reality. Then

I started to teach like that and make everything much more hands-on.

Arduino came about because we wanted to iterate quickly. We wanted something that would be cheap to deploy. We wanted something that a student could use to make a circuit. I also wanted open source software because I didn't want people to have to pay.

So Arduino has an IDE that's cross platform. We had a board that was easy to assemble. We had a little bit of documentation.

[At this point Massimo interrupted the conversation to point out a Fiat prototype on the road. It was covered in cloth so no one could see its shape.]

**DD: How did the team come together?**

**MB:** Arduino started as Wiring, an electronics prototyping platform, which a student of mine was working on. When it was finished, I was committed to using it in school because the only way to try it out was to get it in front of people. Wiring had a more expensive board, about $100, because it used an expensive chip. I didn't like that, and the student developer and I disagreed.

I decided that we could make an open source version of Wiring, starting from scratch. I asked Gianluca Martino [now one of the five Arduino partners] to help me manufacture the first prototypes, the first boards.

Then I invited Tom Igoe from New York University's Interactive Telecommunications Program to come visit Ivrea and work on a project. I had met him and liked him. So he came over, and we ended up using the first Arduinos [to design interactive lamps for Artemide]. Tom liked the concept. He said he'd bring it back to New York and start playing with it.

The idea was to make a board with the minimum number of parts, a PIC processor that would be cheap. I wanted them to cost $20 a board. That's the price of a pizza dinner. So a student could afford to skip pizza and spend the money on a board.

The first run of pre-assembled boards was 200. Fifty were bought by Ivrea. Fifty were

# Dieter Rams
## 10 PRINCIPLES FOR GOOD DESIGN

**Good design is innovative.**

**Good design makes a product useful.**

**Good design is aesthetic.**

**Good design makes a product understandable.**

**Good design is unobtrusive.**

**Good design is honest.**

**Good design is long-lasting.**

**Good design is thorough, down to the last detail.**

**Good design is environmentally friendly.**

**Good design is as little design as possible.**

vitsoe.com/good-design

bought by Sweden. The other 100, we said that we hoped we could sell them. And we sold them. From then on, we had people asking us for boards. When I started to see what people were doing, I knew that Arduino was going to make a difference. ◪

Dale Dougherty is founder and publisher of MAKE.

Anne Brassier / Vitsoe

# The Flora

By *Limor Fried & Phillip Torrone*

## ADAFRUIT'S NEW WEARABLE ELECTRONICS PLATFORM

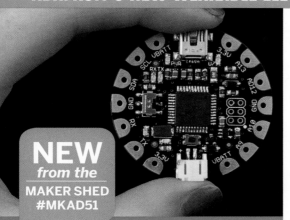

**NEW**
*from the*
**MAKER SHED
#MKAD51**

At Adafruit we're working on a new wearable electronics platform called the Flora. Wearable technology to us is just a temporary term for what's happening with electronics. Practically everyone has an internet-connected supercomputer in their pocket now. It's often stuck to their head, too, with a Bluetooth headset or headphones. It's becoming "wearable," so we think we're bound to see more types of electronics that occupy human real estate.

---

## *Phones augment reality, wearables augment humans.*

---

The Arduino platform has become an easy way to "glue" together ideas, sensors, and applications. In recent years we've been working to make it easier to get sensor information in and out of Arduinos. Wearables are prime for this. We think we'll see an intersection of elegant fashion and thoughtful engineering.

Why? Just look at your phone; it's not just a device for making calls. It's filled with sensors: GPS, proximity, compass, touch, sound, temperature, and more. The smarts in a phone are self-contained, while the sensors within an Arduino-compatible wearable will sense your body and your environment. Phones augment reality, wearables augment humans. Imagine a belt that gently pulls you in the right direction to navigate a city, or an LED jacket that displays the logos and patterns of your choosing.

We started with what we and the open source community wanted in an embedded platform. There wasn't anything out there, so we designed our own: the open source Flora. Here are some of its key features:

» Flora makes it easy to embed LEDs and animations on clothing.
» Comes with projects at launch.
» Includes the Flora-addressable and chainable 4,000mcd RGB LED pixels.
» Has USB HID (Human Interface Device) support, so it can act like a mouse, keyboard, MIDI, etc.
» Modules include: Bluetooth, GPS, 3-axis accelerometer, compass, flex sensor, piezo, IR LED, and more.
» Built-in USB support with Mac, Windows, and Linux.
» Difficult to destroy: the onboard regulator means that even connecting a 9V battery will not result in damage or tears.

Two years ago when we released the Kinect data dump (makezine.com/go/data-dump), we had no idea what would happen, but within months, hundreds, then thousands, of hackers, artists, and scientists made amazing, completely unexpected projects, taking the Kinect to new places. We think that will happen again with the Flora. Available now at makershed.com. ◪

---

Limor Fried is owner and operator of Adafruit Industries (adafruit.com), an open source hardware electronics company based in New York City. Phillip Torrone is creative director of Adafruit and editor at large of MAKE.

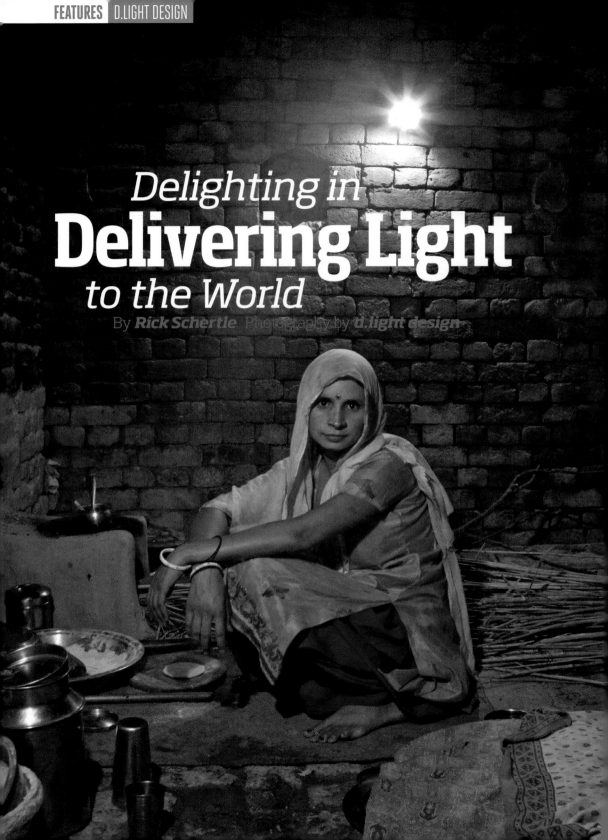

# Delighting in
# Delivering Light
## to the World

By *Rick Schertle*  Photography by *d.light design*

**Like many young, idealistic, and smart "tech" folks, Sam Goldman and Ned Tozun were told over and over that their idea would never fly.**

But their drive, passion to succeed, and bit of naivety beat the odds and now d.light design (dlightdesign.com) is improving the lives of millions.

Currently there are 1.6 billion people in the world without access to electricity. Millions are still using unsafe and inefficient kerosene lanterns. As a Peace Corps volunteer in Benin, West Africa, Goldman experienced the dangers of kerosene firsthand, when a neighbor's kid was severely burned. Together, Goldman and Tozun imagined a world where outdated kerosene was replaced by clean, safe, bright light.

Goldman and Tozun met about six years ago at Stanford University and shared a vision that would eventually become d.light design in 2007. Both passed up lucrative careers with some high tech giants. As founders of d.light, their design challenge was to develop an efficient, affordable, and durable lantern for customers living in some of the poorest regions of the world. The result was a flexible-use solar lantern that can run many hours on a day's solar charge. Their newest design even incorporates a cellphone charger, knowing cellphones are the lifeline of many small business owners in developing countries.

Their challenges include distributing to hard-to-reach markets in over 40 countries and convincing buyers their investment pays off in the end. For farmers and shopkeepers, the investment pays for itself in very little time in the form of fuel cost savings and increased income by being able to work in the evenings.

While the camping industry is potentially a huge market, and their lanterns can be purchased online in the U.S., that's not what they're all about.

With over six million of their lights in use around the world and millions more in the works, Tozun says it's not the money that drives them. With a combination of venture capital and social enterprise (weighted toward social), the money is necessary, but not what keeps the passion going.

Tozun is a totally unassuming guy, even though d.light has received numerous awards, including Forbes Magazine's Top 30 Social Entrepreneurs, and their S250 lamp is featured in the British Museum's History of the

**SOLAR SOLDIERS:** (Above) Sam Goldman (left) and Ned Tozun (right) proudly display d.light lanterns. (Opposite) A woman in India uses a d.light lantern to light her kitchen.

World in 100 Objects as the 100th object. Tozun and his wife worked with d.light in China for three and a half years, and he said he enjoyed the anonymity there.

Tozun's advice to young makers: "Don't give up!" For d.light there were many challenges, but each was met with passion, conviction, and often-naive optimism. Now, with the taste of success, their goals are no longer modest: they aim to improve the lives of 100 million people by 2020. ◪

MAKE contributing writer Rick Schertle teaches middle school in San Jose, Calif. With his wife and kids, he loves all things that fly.

Dezso Molnar fires up the
original Molnari GT gyrocycle.

# The Buckaroo BANZAI of Flight

Written by **Steven Kotler**

## DEZSO MOLNAR ELEVATES FLYING MOTORCYCLES FROM FICTION TO FACT

Photographed by **Cody Pickens**

**The 14,000-square-foot Calfee Design factory is perched on the edge of a bluff in La Selva Beach, Calif. Below the bluff, the Pacific rumbles and moans. Above it, in the shop, on most days they build bicycles. Some of the best in the world, in fact. But today is not most days.**

Today is October 20, 2005. A man named Dezso (pronounced DEZH-ur) Molnar is pushing a strange, four-wheeled contraption out of the warehouse and onto a 2,000-foot runway. A few years back, when Molnar was hunting for a place to build this machine, he had three key needs. The first was isolation. His skunkworks project was the kind of build that attracted all sorts of unwanted attention. Calfee's warehouse fit the bill. It sits on 379 acres of private land and sees few visitors. His second need was expertise. Molnar's contraption had to be light — very light. Calfee's bicycles are made from carbon fiber. They weigh about 14 pounds. Calfee understood light. Molnar's last requirement was a straight stretch of pavement. It didn't have to be a runway, but considering the nature of Molnar's invention, it was a fitting touch.

The true nature of Molnar's invention is hard to discern at a glance. It looks like some *Mad Max* version of a recumbent bicycle, only with training wheels, a chromoly steel roll cage, and a 68-inch, 3-blade propeller cutting through the middle. Today is the very first day Molnar is going to fire up that propeller and see if it can push his machine down the road. He's hoping for speeds about 50mph because, at least according to Molnar's calculations, that's about what it should take to get his flying motorcycle off the ground.

## THE DREAM OF FLIGHT

The flying car and the flying motorcycle are the stuff of dreams — very old dreams. Aviation pioneer Glenn Curtiss invented the

# Flights of Fancy

The dream of flying cars has been around for at least as long as the first airplanes.

1917

1947

Nov. 1947

1949–1956

2012

(Clockwise from top left) The first flying car, Glenn Curtiss' 40-foot-long tri-winged beast made from aluminum (1917); the ConvAirCar, before and after an unscheduled landing (1947); the Aerocar, perhaps the most famous of roadworthy aircraft (1949–1956); and the folding-wing Terrafugia Transition (2012).

first flying car back in 1917: a 40-foot-long tri-winged beast made from aluminum. The beast never did fly, but it managed to hop. That hop was enough, inspiring almost a century of innovation. Next came Waldo Waterman's 1937 winged Studebaker — dead for lack of funding. A bad crash destroyed the 1947 ConvAirCar. The Aerocar went through six iterations before the oil crisis of the 1970s killed off production plans. Since then, there have been dozens of other attempts; a few have flown, most have not.

Today, the two most widely known versions are Paul Moller's M400 Skycar and the Terrafugia Transition. Both of these vehicles are currently for sale; neither of them have actually been delivered to a customer. And that's really the issue. Out of 104 roadworthy aircraft (80 of which have patents on file), none have seen mass production.

There are, of course, good reasons for this. While the upside of a flying car is easy to imagine — no traffic jams, shorter commutes, another excuse to quote *Blade Runner* — the downsides are considerable. Cost and noise, for starters (at $279,000, the Transition has been branded a rich man's toy, to say nothing of Skycar's initial $3.5 million price). But safety and ease-of-use are the bigger stumbling blocks. As with anything that flies, the consequences of pilot error can be severe, especially in bad weather. The safest pilots are the ones with the most practice and the best knowledge of their airplane, twin requirements that further put the flying car out of reach of the average citizen. Moreover, right now, most small planes are high-maintenance, requiring constant (and expensive) upkeep. They also tend to be gas guzzlers. Flying cars, if they're to become become everyman tools, can be neither. And this list doesn't include the bevy of concerns that arrive when one wants to create a street-legal aircraft.

# The Practical Dreamer

Until about eight years ago, Molnar had never intended to get into the street-legal aircraft business, but considering the nature of his pedigree, perhaps it was inevitable. Molnar flew hot air balloons as a teenager, then paid his way through college by flying planes in the Air Force and moonlighting at Truax Engineering, where Robert Truax had a Navy contract for building a replacement vehicle for the space shuttle (this was right after the *Challenger* crash). They built a workable rocket (they were awarded the contract), but funding issues shut down the effort. Afterward, Molnar spent a few years playing music in bands, building robots with the machine performance art outfit Survival Research Labs, and designing DIY vehicles like his buzz-bomb jet-powered go-kart. He next signed on as a crew chief for Craig Breedlove's attempt to drive a jet car through the sound barrier. When that project ran its course, Molnar jumped back into music, and that's where he might have stayed had it not been for London's 2003 heat wave.

"In 2003, I was in the U.K. shooting a music video. It was hot and fun until I got back to Los Angeles, where it was foggy and dew was dripping from the walls in my house. I called a friend and suggested we drive out to Palm Springs, just to warm up. But it was the middle of the day and my friend worked downtown, and the traffic there was bumper-to-bumper. We were trapped. We couldn't leave."

And that's when Molnar got curious about the kind of vehicle that could beat this traffic. He wasn't interested in fairy tales, he was interested in practicality. A flying car capable of vertical takeoff was the most common response, but the only thing that could take off vertically was a helicopter, and those were both expensive and difficult to pilot. But what if he threw that requirement out the window? There are 14,000 airports in America, 30 in the L.A. area alone. What if you could depart from those airports (which typically sit in less congested areas, so getting there isn't as much of an issue) and land in a congested area? There are dozens of parking garages in downtown L.A. — what if you could land atop one of those?

Then Molnar remembered an advertisement from his childhood for a gyrocopter, a type of "rotorcraft" invented in the 1920s by Spanish engineer Juan de la Cierva. Gyrocopters use an unpowered rotor for lift (like a helicopter in autorotation), an engine-powered propeller for thrust (like a plane), and have the advantage of being able to land at very slow speeds (to maximize pilot safety) in extremely small spaces (like the roof of a parking garage). Even better, gyroplanes are cheap — kits start at 10,000 bucks — and easy to fly: a sport pilot's license is required, with just 20 hours of flight time. There was one small issue, however: gyroplanes had a bad habit of crashing.

"The problem," says Molnar, "is that the most popular gyroplane on the market was designed without a horizontal stabilizer — which is what keeps a plane's nose from pitching up or down. People kept getting killed

(Top) Molnar was crew chief for Craig Breedlove during his attempt to break the sound barrier in a car in 1992. (Bottom) Molnar takes his buzz-bomb jet-powered go-kart for a spin.

(Right) Plugs for casting composite body panels for the GT. (Below) The geometry of a street-legal gyrocycle guides the progression of the GT's fairing design.

CABIN

TAIL

because of it, and too often it was considered pilot error. The gyroplane was originally designed to create a safe wing that would not stall, but the perceived option of removing the horizontal stabilizer was a design error. If you put the stabilizer back on, the result is potentially one of the world's safest aircraft — and one that can land in less than 20 feet."

Molnar had solved only part of the problem. He could land in a congested area, but he still had to evade traffic on his way out. "I took a very realistic approach to this question. Since vertical takeoff is too limiting, and no one's going to build a long runway in places like downtown Los Angeles, then flying away isn't the solution. You need to be able to drive. This is where the motorcycle comes in. If you drive away on a motorcycle you can split lanes. It's the fastest way to get away from traffic, and it's legal in 25 countries."

What made the motorcycle even more interesting was the engine. Motorcycle engines are cheap (a new one costs around $2,000 versus $36,000 for most aircraft engines), powerful, durable, quiet, get great gas mileage, and — their best feature — standardized. "Because these engines always have to fit between the driver's legs," says Molnar, "manufacturers devote tremendous time and energy to keep them within racing-restricted displacements (up to 999ccm), yet make them increasingly more powerful. Motorcycle engines are constantly getting better, and you can get them repaired at any roadside shop. It's a perfect solution."

All of these ideas came together in the vehicle Molnar wheeled onto the tarmac that October 2005 afternoon, the Molnari GT gyrocycle. The test drive went exactly according to plan. The one-cylinder engine and propeller generated more than enough wind to push the vehicle past the 50mph mark. Road tests confirmed it could master the freeways: the two-wheeled bike did 90mph no problem. (The newer, tri-wheeled G2 does 160mph and handles like a sports car.) Because of his engine choice, Molnar also evaded a problem that plagued Moller and Terrafugia: his vehicle passed a smog test. The bike was street legal. Now it was time to see if it could fly.

Dezso Molnar and Craig Calfee prepare the latest version of the gyrocycle, the Molnari G2, for upcoming road testing at Bonneville Salt Flats, Utah.

## DREAMS REALIZED

The first flight took place in 2006, and everything went perfectly. The crew logged four hours of flight time without a mishap, so Molnar started thinking about production. Originally, his plan was to sell single-seaters, constructed from kits. But his backers wanted to explore upscale markets — which meant two seats, room for cargo, and all kinds of doodads. So Molnar and his partners spent years working on the new G2 while trying to convince the FAA that gyroplanes could be engineered safe enough to be sold as turn-key light sport aircraft. The FAA wouldn't budge, so Molnar went back to his roots.

"I decided," he says, "that this had to be a DIY project. The FAA won't allow a ready-made gyroplane, but tens of thousands of aircraft flying are homebuilt, and the FAA regulations allow gyroplanes that are built from kits. I had a machine that could fly and drive, but if it was going to help launch the flying car revolution, then DIYers would need to assemble their own to push it forward."

Unlike kit airplanes, which can take years to build in a hangar, Molnar thinks the G2 could be snapped together in a garage in four weeks. But it hasn't been engineered for production, or optimized. Engineering, he feels, is easy; it could even be crowdsourced. But optimization requires testing — and this is where Molnar's business plan deviates further from his peers. Rather than drum up financing for mass production, Molnar wants to establish a gyrocycle racing league populated entirely by kit builders.

"Sure, it's not a huge market, but this is not a machine for couch potatoes," Molnar says. "It's for skydivers and guys who ride Ninjas. These people live for adventure, they know they're on the cutting edge, and consider what happens if we're successful: suddenly the chuckle factor is gone from the flying car discussion. In its place, you get all the optimization and engineering that results from any race program, and nobody streamlines better than a racing pit crew." ◪

Steven Kotler's (stevenkotler.com) books include *Abundance*, *A Small Furry Prayer*, and the novel *The Angle Quickest for Flight*.

# The $100 Secret Room

By *"CT"*

aka *Crazy Talk*

## Rule number one of having a secret room: don't tell people you have a secret room.

But this is too good not to share, so I guess I'll just leave off my real name so that it's still somewhat secret.

The lady who manages my rental told me that I could do whatever I wanted to the place as long as it was an improvement. While "improvement" can be subjective, who wouldn't want a secret room? (I sometimes imagine my departing argument, where I dare her to post the place on Craigslist, mentioning the secret room, and see how fast it gets rented.) Okay, on to the how and why.

As you can see, the dining room has a door leading to a small office (or whatever else a 10'×10' room can be used for). The doorway is quite wide, and I venture to guess that originally there was no door. As with a lot of the past "improvements" here, someone added a door and covered the remaining space with a pseudo-wall. This had always bothered me, especially because once you enter the room, the light switch is found behind the door. It would be a pain to change the hinges on the door or relocate the light switch, so I opted for a win-win idea.

Just like any kid, I always wanted a secret room. I thought about it for a long time and it became clear that this room was perfect for it, but it had to be done right. Everyone seems to think of the bookcase door first, but this wall just wasn't right for that. Real construction and investment would be required to make that work. Besides the weight factor, I didn't want to have to find things to fill a bookcase, so I went simple and cheap.

The advantages of my simple design are that it makes up for any of the guaranteed inaccuracies in my workmanship and still leaves the door mostly undetectable. (It's also worth mentioning that the walls aren't entirely plumb and the roof is slanted for drainage.)

After removing the door and bad patch job, I had to do some simple framing in the doorway. I then opted for simply skinning the wall with ⅛" lauan plywood. The idea was that I would take advantage of the fact that the area is 10' long by 8' high, for the most part. By making a grid of 2'×2' panels with 1"×3" wood planks, I would accomplish four things:

» Create uniformity that would throw off an observer to the idea that there is a door present.
» Allow the door to overlap the inside of the planked frame to hide the edges.
» Allow for a door measuring 2'×4', which I felt was ideal for the project.
» And again, allow for the trapezoidal wall and inaccurate cuts on my part.

I knew that even if it didn't really work out to be awesomesauce, it would improve the look of the wall.

To make it seem more like a design element, I decided to stain the planks slightly darker than the panels. After skinning the wall, I stained it as well as the door panel. The difficult part was having to pre-cut and number all the 1"×3" planks, then stain them separately so as to not get any dark stain on the lighter panels. To be fair, the stain took the budget just over $100, but it was damn close.

Once all the planks were in place, I fitted the door with hinges and a door-closer I found at Habitat for Humanity on the cheap. It didn't have an arm, so I just made one out of scrap metal stock I had lying around the garage.

Satisfied with the project, I left the backside unfinished. Once you're inside, it really gives it that secret room feel. Besides, I might have to do some deconstruction to get some of my items out of this room when I move. I knew this might be a problem from the beginning, but decided it would be worth it. I finished it off with a framed photo of a duck as a friendly reminder to not bump your head on the way out.

I'm ready for the zombies, are you? ◪

"CT" (jetsetpress@sbcglobal.net) is a freelance Steadicam operator who also tinkers in the garage doing amateur welding, woodworking, aluminum brazing, Vespa restoration, artwork, and the occasional hydrogen fuel cell experiment.

James Grover's rocket blasts off.

# *Unreasonable* Rocketeers

THE NEW SPACE RACE IS MAKER-MADE

Written and photographed by

*Charles Platt*

**While Elon Musk ponders his personal ambition to visit Mars, privately funded space ventures in the grungy little backwater of Mojave, Calif., continue to make rapid progress on a more immediate timescale.**

I wrote about them here two years ago ("Rocket Men" in MAKE Volume 24), but events at Mojave have taken an exciting new twist. The Mojave Air and Space Port now has its very own makerspace called Mojave Makers, and the line between professional engineers and the rest of us is becoming difficult to discern.

Scaled Composites, founded by Burt Rutan, is at the high end of the funding food chain in Mojave. Its most grandiose venture is to build a dual-fuselage aircraft using parts cannibalized from a pair of 747s. It will have the largest wingspan in the world, to carry a SpaceX Falcon 9 rocket for launch from a high altitude, greatly increasing its potential payload.

At the mid-level, Xcor has started selling suborbital flight tickets for rides on Lynx, its home-brewed rocket plane. By the end of 2013 it should be flying to the edge of space.

At the grassroots level, almost anyone can launch his own homemade rocket from a desert facility operated by Friends of Amateur Rocketry, just 20 miles north of Mojave (see FARther Out, page 77).

Traditionally these groups have operated in relative isolation from each other, but the Mojave Makers space promises to bring them together. It will also provide tools and equipment that few individuals could normally afford, and so much floor area, you could build a small airplane in it.

When I attended the launch party earlier this year, I found myself in a big, bare-bones building that lacked amenities but already contained a miscellany of tools. They ranged from a simple drill press to a 30-horsepower Puma 10S lathe weighing 8 tons, enclosed in a metal shell almost the size of an SUV.

In an adjacent room, seats extracted from the first-class cabin of a retired

*Follow us @make* **73**

jetliner had been arrayed to form a social area. A simple attitude-control device and an advanced helicopter concept were on display. Outside, in empty land alongside the building, Lee Valentine of the Space Studies Institute was eyeing the area where he hopes to establish a simulated self-sustaining space habitat (see: Extraterrestrial Life below). Clearly, the makerspace is going to make its mark on Mojave.

## INVASION OF THE CITIZEN SCIENTISTS

Michael Clive was the initiator. Until a few years ago, he lived in a one-bedroom apartment in Venice, Calif., worked for DreamWorks Animation by day, and frequented the Los Angeles club scene by night. Then he got involved with Crash Space in Culver City, became its facility manager, and helped to build it into a premier center for makers. Now he's hoping to achieve the same thing here, when he's not doing his day job at Xcor.

He rejects the word "amateurs" to describe the makers of space hard-ware who have infiltrated a field that used to be populated exclusively by credentialed engineers. "They're citizen scientists," he says. "They're more adventurous than model-rocket hobbyists. They've seen a lack of progress in space technology, and have taken the initiative to advance the state of the art. They're finding engineering solutions to problems that have been abandoned or avoided by the major players, for instance, in small propulsion and intelligent guidance systems."

Four friends collaborated with Clive to establish the facility: Ethan Chew, Scott Nietfeld, Andrew Bingham, and Nadir Bagaveyev, all of them employed at Mojave space companies. Bagaveyev is Russian-born but now presents himself as a hardcore American capitalist. "I want to build rockets that make money," he says. "But in the makerspace, other people are pro-communal. So we head-butt sometimes, but we can have a balance here. It's a hangout communal place, and it's also a place for people who want it to be an incubator. We

# Extraterrestrial Life

Zero-gravity conditions are bad for many bodily organs and functions, but what about lunar gravity, or conditions on Mars? Could we live there and have normal children? No one knows. Nor do we know how to build a space colony that is indefinitely self-sustaining.

This year the Space Studies Institute (SSI) launched its "Great Enterprise Initiative" to find out if people really can live in space or on other worlds. The plan is to build two laboratories: G-Lab, testing the long-term effects of low gravity on humans in orbit, and E-Lab, a closed-loop environment here on Earth.

SSI chairman Lee Valentine hopes to site E-Lab on vacant land right beside the Mojave Makers space.

While life-support systems on the International Space Station pump air through filters and activated charcoal, Valentine wants an environment that is less dependent on mechanical devices that must be maintained, or supplies that must be replenished.

The environment should also yield enough food and water to sustain its residents. "We have ideas about which kinds of plants to use in such a system," he says, "but until we put them together, we don't know that it's going to work. We don't even know how we will introduce energy to the plants. Can we grow enough food with photodiodes?" He wants to find out.

Our terrestrial environment is self-regulating and self-cleansing, partly because organic volatiles are oxidized in the atmosphere. But in space, you don't have as much sunlight, or the atmospheric mass. Nor do we know how much $CO_2$ will be optimal. "All plants grow better in higher levels of $CO_2$ than we have here on Earth," according to Valentine. "We have to look at how much $CO_2$ we need."

He sees the adjacent Mojave Makers space as an ideal place to fabricate components, and since many makers are SSI members, the synergy is obvious. "SSI has always been supported by its associates," he notes, no doubt hoping to tap the enthusiasm of engineers and citizen scientists alike.

Members of Friends of Amateur Rocketry head to the launch pad. (Inset) Lee Valentine of the Space Studies Institute on the empty lot beside the Mojave Makers space, where he hopes to see a test facility for habitable space environments.

*The line between professional engineers and the rest of us is becoming difficult to discern.*

just wrote it in the bylaws that Mojave Makers will never claim property rights to anything developed here. I'm designing a UAV thing in my little corner. As soon as it's mature enough, I will form a company to develop it further."

Scott Nietfeld of Masten Space Systems sees the location as a great way to facilitate networking. "There were no social outlets at Mojave," he says. "When I got here, I knew there were all those people at other companies building interesting things, but I never saw any of them. So I wanted

to start a place where people could work together."

He shows me a beautiful little hardwood case that he made for his Kindle. "I like working with my hands," he says. But space travel remains his primary interest. "Building rocket vehicles is still not easy, but it doesn't require a huge corporate conglomerate anymore. I mean, people now build liquid-fueled rocket engines as a hobby."

The internet has played a part in this. You want to know how NASA designed the guidance systems for the Apollo moon missions? Just check

(Left to right) Michael Clive, prime mover of the Mojave Makers space. Scott Nietfeld, Nadir Bagaveyev, and Andrew Bingham, all cofounders of Mojave Makers. (Below) Getting ready for ignition on the FAR launchpad.

the NASA Technical Reports Server (NTRS), where the archives are publicly accessible.

Unfortunately, as the old saying goes, a government big enough to give you everything you want is big enough to take it all away. The FAA has set altitude limits for classes of amateur rocketry. You can't buy or own some types of propellants without appropriate licenses. And strict federal laws known as International Traffic in Arms Regulations (ITAR) control almost all space hardware.

Andrew Bingham, who attended ITAR briefing sessions when he used to work at the legendary Jet Propulsion Laboratory, gives me an example of the problem. "You can buy Arduinos rated for -25 to +40 centigrade. So if that board will run in a vacuum, it may also work in a spacecraft. But suppose you exchange the electrolytics with solid capacitors. My reading of the rules suggests that it's now designed for the space environment, and consequently it's controlled by ITAR. Can I even talk about it publicly anymore?"

Bingham has some ambitions that are not space-related: "I want to build a solar hot water heater for my spa," he says. "And some custom computer keyboards." But, he continues, "I also want to test some commercial electronics to see if they'll work in a cube sat," which is a 10-centimeter cube that can be launched into orbit. "Typically the kits to do that cost thousands of dollars. I want to do it for $500."

Who could object to a project like that? Well, that remains to be seen. When today's plans turn into tomorrow's functioning devices, perhaps another two years from now, we may learn just how tolerant the government will be of the fiercely independent maker spirit. ◢

Charles Platt is the author of *Make: Electronics*, an introductory guide for all ages. A contributing editor of MAKE, he designs and builds medical prototypes in Arizona.

# FARther Out

Drive 20 miles north of Mojave, take a lonely dirt road across the baked gray sand of the desert, and eventually you find a 10-acre rocket launching facility maintained on an all-volunteer basis by FAR, the Friends of Amateur Rocketry.

Two cinder-block bunkers face some small launch pads and a steel gantry. Behind the bunkers are prefabricated buildings for vehicle assembly, repairs, and modifications, with tools including a lathe, mill, drill press, chop saw, grinder, and welder. There's also a microwave oven, bottled water, and basic food supplies (restrooms are still under construction).

Rocketeers converge here every other Saturday, driving pickup trucks and station wagons loaded with rockets, spare parts, and tools. Today the Camarillo High School rocket club is assembling a vehicle that should reach a speed of Mach 1.8. Nearby I find Paul Breed, a hardcore individualist and uber-maker who founded his own company, aptly named Unreasonable Rocket, to build a viable contender for the Northrop Grumman Lunar Lander X Prize Challenge. He did it all with the collaboration of just one person: his son.

Today Breed is at FAR testing a parachute system, but he says he's also in the desert for fun. "There's a good chance to watch fireballs and destruc- tion," he says with a grin. "Energetic chemicals don't always behave." Still, his ambitions are serious. He hopes to compete in the $3 million Nano-Satel- lite Launch Challenge. "Getting to orbit is really hard," he says reflectively. "I'm thinking of using a three-stage rocket."

A more low-key enthusiast, James Grover is an engineer at Northrop Grumman who specializes in CAD design and worked on the James Webb space telescope. He comes to FAR for the pleasure of hands-on work. "This gives me an outlet to do exactly what I want," he says, adding that for him, rockets are a lifelong passion. He sets up and fires his 7-foot rocket, then disap- pears out across the desert in his pickup truck, using radio equipment to find where it came down. Later he reports that it reached more than 42,000 feet. His next project will be a two-stager that should exceed 100,000 feet.

I ask one of the site managers what I'd have to do to fly my own rocket out here. "Probably show up with it," he says laconically. Actually some paper- work is involved, to comply with FAR's state, federal, and local licenses. Even in this wide-open empty place, there are regulations. But getting access couldn't be much more affordable. You can participate for a mere $10 admission fee.

Friends of Amateur Rocketry lives up to its name (left). The observation bunkers at FAR, with mission control table (right).

## RESOURCES

FAR: friendsofamateurrocketry.org
Scaled Composites: scaled.com
Xcor: xcor.com
Mojave Makers: mojavemakers.org

# The Island of Magnificent Toys

Written and photographed by **Gregory Hayes**

Toy sculptor Scott Hensey in our new video series, *Make: Believe*.

"Oh, YOU made that?" We must have said it a dozen times during our recent visit to Anaglyph Sculpture, where every surface is crowded with decades worth of toys and collectibles produced by sculptor Scott Hensey. Almost anyone would recognize a few favorite commemorative items, but we geeked out over the Star Wars phones, Happy Meal toys, Disney collectibles, and thousands of colorful items that inspired us to play and thus begin our own creative journeys.

By chance, Hensey's knickknack wonderland is right down the street here in Sebastopol, Calif., making it a perfect spot to kick off MAKE's new video series, *Make: Believe*.

We're taking our cameras into the studios of the makers who turn fantasy into memorable reality. Whether iconic or incognito, they design and make the creatures, props, special effects, toys, games, and, often, the very fabric of modern culture. They stretch our imaginations and change our whole concept of what's possible. We're going to show you how they work their magic.

Here's a peek at what we saw while Hensey talked about technique, inspiration, getting started in a tough business, and how to make a living doing what you love. You can see our video and all of the photos from our visit online at makezine.com/go/believe.

Besides wrangling MAKE photography, Greg helps produce *Make: Believe*.

# The Case for Good Design

## Monster MIDI

If you think all MIDI controllers look like synths, think again. Tristan Shone's hardcore industrial fetish yields handmade Drone Machine controllers in unlikely forms, like the 300lb disc Rotary Encoder and the Linear Actuator, a weighted slide handle with spring-loaded plungers at either end. Learn how to make the Headgear controller in Volume 22 and check out his sights and deep, dark sounds at authorandpunisher.com.

—*Goli Mohammadi*

Jeni Cheung (Tristan Shone): Benjamin Cowden (Shaker)

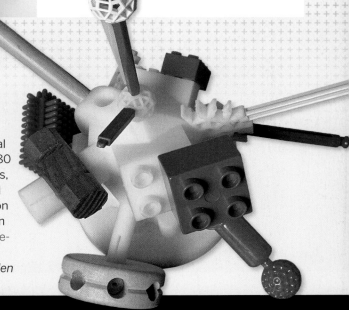

## MAGICAL SOLDERING UNICORN

Following a conversation on L.A.'s Crash Space mailing list about how hackerspaces don't often attract girls, a joke was made about a soldering unicorn. (Crash Space actually bucks the trend, with a 50/50 gender split.) The idea culminated in Sparkles, a My Little Pony figurine fitted with a working soldering gun in place of its horn. Crash Space member Sean Bonner notes, "We actually use her all the time for hands-free soldering help." makezine.com/go/sparkles

—Laura Cochrane

# Block Party

Break down the walled gardens of kid construction sets with the Free Universal Construction Kit. The full set of almost 80 pieces connects 10 popular building toys, including Lego, K'Nex, Lincoln Logs, and Tinkertoy. Now your Lincoln Log mansion can have that Lego laser defense system you've been dreaming about. fffff.at/free-universal-construction-kit

—Craig Couden

**Want to mix up some mudslingers without freezing your fingers?**

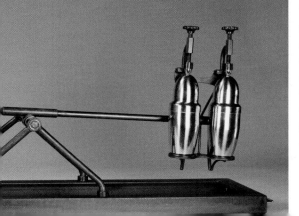

# Warm Hand Shake

Wind up the Post-Imperial Shaker, which mechanically agitates two cocktail shakers driven by a hand crank. Artist Benjamin Cowden's ruggedly handsome machine is also elegantly useful; according to his website, "The ingredients are swirled horizontally as well as vertically." It's fascinating to watch, before or after enjoying the results.

twentysevengears.com        —Gregory Hayes

# Ambarometer

I've always had a fascination with wordless user interfaces. Visual metaphors are, in effect, a higher-bandwidth device-to-brain link, using the brain's built-in pattern-recognition circuitry. And when implemented properly, the results are always visually pleasing. Jonathan Foote clearly used these principles to design his elegant and ingeniously functional Ambarometer. This device does triple duty as barometer, ambient lamp, and sculpture. It's hard to appreciate the Ambarometer on the merits of the photograph alone, but there's a video, viewable at rotormind.com.

*—Max Eliaser*

# Big*Deal

Style is in the eye of the beholder with this rad, 80s throwback iPhone 4/4S case. Built from four 3D-printed pieces, feed in your iPhone earphones and you're ready to wheel and deal like Gordon Gekko. Download the specs for free, or print it from Shapeways and they'll plate it in gold! fffff.at/brickiphone

*—CC*

# Eco Print & Build

Researchers from the College of Environmental Design at UC Berkeley have developed methods for 3D printing inexpensive yet sturdy materials for use in architectural applications. The team's hacked CAD/CAM machines print tough, durable parts out of a homemade, translucent, cement polymer. Complex or large designs, like the Seat Slug, are printed in pieces and then assembled. rael-sanfratello.com/?p=1154

—*Josie Rushton*

# Toolbox
## BOOMBOX

**Maker Floyd Davis IV creates booming sound systems in everyday items. The customization options are limitless. Build your own amplifier, add a rocking LED light system, or just trick it out with some extra bling. A totally unexpected way to liven up your next social gathering with some funky beats.** artpentry.com

—*Ben Lancaster*

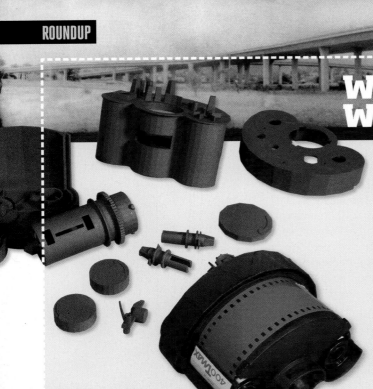

# Widescreen Wonder

Eben Ostby fell in love with an old camera called a Kodak Panoram, which captures flat swaths of landscape in a panoramic sweep. But he wanted to use 35-millimeter film. So he drew up a prototype on SketchUp, and sent his design off to 3D printing company Shapeways. With the pieces assembled, he rigged a piano-wire spring that rotates the lens. See his wistful, contemplative shots at ebenostby.com.

—Bob Parks

**"Hacking is fun. Don't worry about finding a reason why."**

## THE MAKEYS
### Maker Hero 2012

# Queen OF HACKS

To self-taught electrical engineer Jeri Ellsworth, deconstructing and re-engineering the world around her is second nature. This curious and enthusiastic approach to life shines through in her numerous hacks, which include a brainwave-activated "aha!" light bulb head accessory, a microphone made of razor blades and Tupperware, a dress that lights up when people approach, and a Commodore 64 that she hacked into a bass keytar. Ellsworth's designs are hybrids with humor. As she says on her YouTube channel, "Hacking is fun. Don't worry about finding a reason why."

makezine.com/go/ellsworth

—LC

Eben Ostby (Kodak Panoram);
Bridgette Vanderlaan (Ellsworth)

# Enticing Enclosures

Thingiverse is a site where folks can share digital designs for laser cutters, CNC machines, and (mostly) 3D printers. Looking for a slick enclosure for your next project? Check out the collective knowledge of the maker community.

Adbo Arduino Box (oomlout)
thingiverse.com/thing:6940

Revolver iPhone (juniortan)
thingiverse.com/thing:3241

Adafruit Beagle Bone Box
thingiverse.com/thing:18632

Ice Tube Clock Enclosure (nmatsuda)
thingiverse.com/thing:3124

Gameduino Acrylic Case (skpang)
thingiverse.com/thing:9883

Raspberry Super-Pi Case (RichRap)
thingiverse.com/thing:25363

Reddit Upvote/Downvote Button
(TheNewHobbyist)
thingiverse.com/thing:10423

Shruti-1 Synth Enclosure
(watsdesign)
thingiverse.com/thing:16629

# SKILL BUILDER

INTERMEDIATE

# Get Started with
# BeagleBone

## Got a project too big for a microcontroller? This embedded Linux board offers powerful features in a small package.

Written and photographed by *Matt Richardson*

**MANY MAKERS LOVE MICROCONTROLLER** platforms like the Arduino, but as the complexity increases in an electronics project, sometimes a microcontroller just won't cut it and you need something with a little more "oomph."

For example, if you want to use a camera and computer vision to detect dirty dishes in your sink, it might be a good idea to explore your options with embedded Linux platforms. These boards are generally more powerful and capable, and are sometimes the perfect solution for projects that are too complex for our beloved microcontrollers.

Not only that, but as the price of embedded Linux platforms drops, the community of support around them grows, which makes them much more accessible to novice and intermediate makers than ever before.

The BeagleBone is an embedded Linux development board that's aimed at hackers and tinkerers. It's a smaller, more bare-bones version of the BeagleBoard. Both are open source hardware and use Texas Instruments' OMAP processors, which are designed for low-power mobile devices.

These days, a typical microcontroller-based board costs $20 to $30, while the BeagleBone retails for $89. Other than a more powerful processor, what are you getting for your extra money?

**» Built-in networking:** Not only does the BeagleBone have an on-board Ethernet connection, but all the basic networking tools that come packaged with Linux are available. You can use services like FTP, Telnet, and SSH, or even host your own web server on the board.

**» Remote access:** Because of its built-in network services, the BeagleBone makes it much easier to access electronics projects remotely over the internet. For example, if you have a data-logging project, you can download the saved data using an FTP client or you can even have your project email you data automatically. Remote access also allows you to log into the device to update the code.

**» Timekeeping:** Without extra hardware, the board can keep track of the date and time of day, and it's updated by pinging internet time servers, ensuring that it's always accurate.

**» File system:** Just like our computers, embedded Linux platforms have a built-in file system, so storing, organizing, and retrieving data is a fairly trivial matter.

**» Multiple programming languages:** You can write your custom code in almost any language you're most comfortable with: C, C++, Python, Perl, Ruby, Java, or even a shell script.

**» Linux software:** Much of the Linux software that's already out there can be run on the BeagleBone. When I needed to access a USB webcam for one of my projects, I simply downloaded and compiled an open source command-line program that let me save webcam images as JPEG files.

**» Linux support:** There's no shortage of Linux support information on the web, and community help sites like stackoverflow.com

# MATERIALS & TOOLS

**» BeagleBone development board** item #MKCCE1 at Maker Shed (makershed.com)
**» Power supply, 5V**
**» Breadboard**
**» Jumper wires**
**» LED**
**» Resistor, 50Ω–100Ω (1) and 10kΩ (1)**
**» Switch, momentary pushbutton**
**» Computer with internet connection**
**» Router and Ethernet cable**
**REQUIRED RESOURCES (ONLINE):**
**» Ångström distribution of Linux, latest version:** beagleboard.org
**» BeagleBone's System Reference Manual:** beagleboard.org/bone
**» mrBBIO Python module:** github.com/mrichardson23/mrBBIO
**» PuTTY, an SSH/Telnet client (optional)** if you're using a Windows PC: putty.org

come in handy when a challenge comes along.

**» Multitasking:** Unlike a basic microcontroller, embedded Linux platforms can share the processor between concurrently running programs and tasks. For example, if your project needs to upload a large file to a server, it doesn't need to stop its other functions to wait for the upload.

**» USB:** The BeagleBone can act as both a USB host and a USB device — not only can you control it from your computer, you can also connect USB devices to it. This makes it easy to integrate common USB peripherals like flash drives, wi-fi adapters, and webcams into your projects.

**» Size:** The BeagleBone packs all these features into a small form factor. In fact, it fits perfectly into an Altoids tin!

Even though these platforms are becoming easier to work with, it helps to be well versed in digital input and output (I/O) before tackling embedded Linux for your physical computing projects. Arduino is a great platform for getting started with GPIO (General Purpose Input/Output); to learn more, visit makezine.com/arduino.

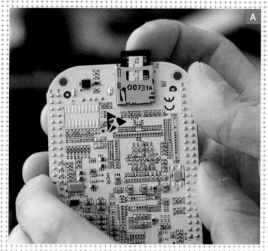

```
Terminal — ssh — 80×24
Last login: Thu Aug 30 06:50:25 on ttys000
Matt-Richardsons-MacBook-Pro:~ matt$ ssh root@beaglebone.local
The authenticity of host 'beaglebone.local (10.0.1.86)' can't be established.
RSA key fingerprint is e0:41:05:f8:54:cb:32:b1:62:a6:bd:01:88:73:aa:b8.
Are you sure you want to continue connecting (yes/no)? yes
Warning: Permanently added 'beaglebone.local' (RSA) to the list of known hosts.
root@beaglebone.local's password:
root@beaglebone:~#
```

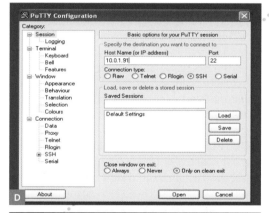

## Getting Set Up via Ethernet

Your BeagleBone comes with a MicroSD card preloaded with a customized version of the Ångström distribution of Linux. Since development on this distribution happens rapidly, you'll want to update to the latest version, available at beagleboard.org.

To access the BeagleBone to upload code, you can connect to it over the Ethernet port with SSH, or you can connect it directly to your computer's USB port. Since we'll be downloading a few files onto the board directly from the internet, let's connect to the BeagleBone via Ethernet.

**1.** With the MicroSD card inserted **(Figure A)**, connect the BeagleBone to your router via Ethernet and plug in a 5V power supply to the BeagleBone **(Figure B)**.

**2.** On a Mac or Linux box, open a terminal window and type `ssh root@beaglebone.local` **(Figure C)**.

On a Windows PC, download PuTTY and open it. Enter `beaglebone.local` as the host address, making sure the SSH button is selected, and press Open. When it shows you the prompt `login as:`, type `root` and press Enter. If the address `beaglebone.local` doesn't work, try using the IP address of the board instead **(Figure D)**.

TIP Find your BeagleBone's IP address by logging into your router and looking for "beaglebone" on the DHCP clients list **(Figure E)**.

**F**

**3.** The first time you connect, your SSH client may warn you that the host is unknown **(Figure F)**. It's OK to dismiss this message.

**4.** There's no password by default, so just hit Enter. You know you're connected when you see the `root@beaglebone:~#` prompt.

## Controlling Pins from the Command Line

Before we get into writing code, let's look at how to do basic digital pin control from the Linux command line. Once we understand how the Linux kernel uses a virtual file system to read and write pins, it makes programming the BeagleBone much easier. (It's also possible to read and write specific memory registers to access the pins, but that method is more advanced.)

### 1. OUTPUT VIA GPIO PINS: LIGHT AN LED

A great way to get to know a new platform is simply getting an LED to light up, so let's wire up an LED.

**1a.** The BeagleBone has 2 main sets of headers, each with 46 pins. One header is labeled "P8" and the other is labeled "P9." Only the end pins are labeled, so you'll have to count pins from the ends to determine the pin you want to access. Put a jumper wire in one of the ground pins, which are pins 1 and 2 of header P8 and P9.

We'll connect our LED to pin 12 on header P8. Put another jumper in that pin, counting off even numbers from pin 2, as shown in **Figure G**.

# DERIVING THE LINUX GPIO SIGNAL NUMBER

» The GPIO signal numbers you'll refer to within the Linux file system are not the same as the pin numbers printed on the board. Here's how to derive the Linux GPIO signal number from the hardware pin number:

**1.** Download the BeagleBone's System Reference Manual from beagleboard.org/bone.

**2.** In the section of the System Reference Manual that shows the pinouts for P8, you can see that the default signal name for hardware pin 12 is GPIO1_12. (The signal names take the format of GPIO*chip_pin*.)

**3.** To determine the pin number that you'll use within Linux, multiply the chip number by 32 and add the pin number. So for signal GPIO1_12, we'll be referring to it as GPIO signal 44. (32 × 1 + 12 = 44.)

**4.** It's important to know that many pins can be assigned different functions, not just digital input and output. This feature is known as pin multiplexing or "pin muxing" and it can make things a little tricky.

For this tutorial, I'm using pins that default to GPIO mode when the BeagleBone is powered on. Many of the pins default to other modes — and the System Reference Manual doesn't always reflect these defaults correctly. The defaults can also change as updated versions of Ångström are released for the board. This can make the process a little tedious, but later in this tutorial, I'll introduce you to a Python module that will take care of the mux setting automatically.

**1b.** On a breadboard, connect the cathode (−) of an LED to ground and the anode (+) to pin 12 on header P8 through a current-limiting resistor (any value between 50Ω and 100Ω should do).

**1c.** Figure out the Linux GPIO signal number for pin 12 on P8 (see sidebar, page 90).

**1d.** Now that we know which pin number to use within Linux and we've set it to GPIO mode (pin 12 defaults to GPIO mode), let's use the command line to control the pin.

On the command line, change to the *gpio* directory **(Figure H)**:

```
cd /sys/class/gpio
```

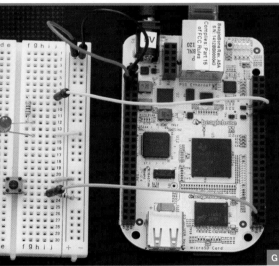

G

H

**1e.** When you list the contents of the directory with the command `ls` you'll notice there's no folder for GPIO signal 44. That's because first we need to export the pin to "user space" so that we can control it. To do that, write the number **44** to the *export* file:

```
echo 44 > export
```

**1f.** Now when you type `ls` you'll see the directory *gpio44*. Change to that directory:

```
cd gpio44
```

**1g.** Since we're trying to control an LED, we'll need to set the pin as an output by writing the word **out** to pin 44's *direction* file:

```
echo out > direction
```

**1h.** Now we're ready to set the pin high to illuminate the LED. Write 1 to the *value* file:

```
echo 1 > value
```

**1i.** Naturally, setting the pin low and turning off the LED means writing a 0 to the *value* file:

```
echo 0 > value
```

## 2. INPUT VIA GPIO PINS: READ A BUTTON

Using the GPIO pins as an input is just as easy. Here's how to tell the BeagleBone to read a pushbutton switch.

**2a.** Wire up a momentary pushbutton to pin 45 on header P8 with a 10K pull-down resistor. Connect the other side of the button to 3.3V source on header P9 pins 3 or 4 **(Figure G)**.

**2b.** First, we must export the pin to the user space and change to its directory. Since pin 45 on header P8 is **GPIO2_6**, we'll export pin **gpio70**:

```
echo 70 > /sys/class/gpio/export
cd /sys/class/gpio/gpio70
```

**2c.** Set the pin direction as an input:

```
echo in > direction
```

**2d.** Now instead of writing the *value* file, we'll read it:

```
cat value
```

**2e.** This should return 0 for a low pin. Now press and hold the button while you execute the `cat value` command again. If you have the button wired up correctly, you should now see a 1, indicating the pin is high.

**2f.** When you're done with the pins, be sure to unexport them from the userspace:

```
echo 44 > /sys/class/gpio/unexport
echo 70 > /sys/class/gpio/unexport
```

## Using Python to Control the Pins

As you can see, using pins as digital inputs and outputs is as simple as reading and writing files in the Linux Virtual File System. This means that, without any libraries, you can use any language you're comfortable with, as long as there's a compiler or interpreter for that language available on the BeagleBone.

The included Ångström distribution of Linux even includes a built-in web-based development environment for Node.js called Cloud9. There's a framework called Bonescript currently under development which can be used for accessing GPIO pins within Node.js.

However, when I started my first few programs with the BeagleBone, I decided to use Python because I was more comfortable working in Python than in Node. In my first Python script, I was working with the files manually: opening them, reading or writing them, then closing them each time I wanted to read or write a pin.

This became tedious, so I wrote a Python module called mrBBIO, which packages up all those functions into an Arduino-like syntax. It also lets you refer to the pins on the BeagleBone as their physical pin locations, so you don't need to refer to the System Reference Manual to determine the Linux signal name for the physical pin or figure out how to change its mux setting. (I was inspired by Alexander Hiam's pyBBIO module, which instead writes to specific memory registers to control the pins.)

As long as your BeagleBone is connected to the internet, you can download mrBBIO directly from GitHub. To do so, first change into your home directory:

```
cd ~
```

and then download the latest version of mrBBIO:

```
git clone git://github.com/
mrichardson23/mrBBIO.git
```

This will create a directory called *mrBBIO*. Change to that directory:

```
cd mrBBIO
```

If you review the example code (type `cat example.py` to see it), you'll see that it has `setup` and `loop` functions. Just like Arduino, the `setup` function runs once when the code is first executed and then the `loop` function runs repeatedly until the program is terminated. The `setup` function in this example sets pin P8.12 as an output and P8.45 as an input:

```
def setup():
    pinMode("P8.12", OUTPUT)
    pinMode("P8.45", INPUT)
```

The `loop` function will be checking whether the button is pressed. When it senses that it was pressed, it will turn the LED on for 1 second and then turn it off. It will also output text to the console to indicate when the button was pressed, using an Arduino-like `millis()` function.

To execute the code from the command line, type:

```
python example.py
```

and watch the LED light up when you press the button! To exit the program, type Ctrl-C. The mrBBIO module will take care of unexporting the pins for you.

If you're eager to start experimenting on your own, you can start by using the example file as a template. Make a copy of the file:

```
cp example.py test.py
```

and edit it in Nano (or your preferred text editor):

```
nano test.py
```

If you'd like, you can even use your computer's text editor and upload the code to your BeagleBone via SFTP.

## Taking It Further

Of course, using an embedded Linux system to blink an LED is overkill, but this guide will give you the basic tools you need in order

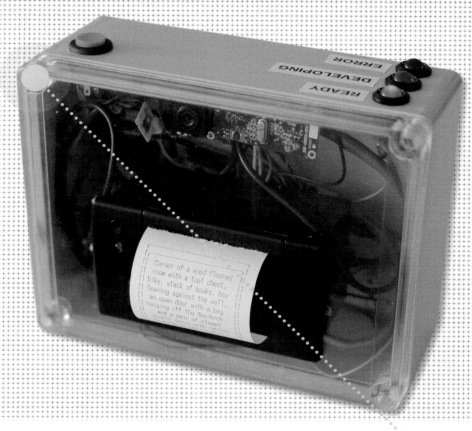

to take on more complex projects. With all the extra features that the embedded Linux affords, there's so much you can do with these boards and some simple GPIO. You could create an internet-connected coffee pot that serves its own web-based control panel for activating it and changing its settings. Or perhaps you'll want to log temperature data directly to a text file on a USB flash drive.

I recently used the BeagleBone for my Descriptive Camera project (mattrichardson. com/Descriptive-Camera). It's a camera that outputs a text description of the scene you capture instead of a photograph **(Figure I)**. It accomplishes this through crowdsourcing. When the shutter button is pressed, the camera snaps a picture from a USB webcam and then contacts my friends via instant message to see if they can describe the image. When they respond, their response is printed out on a thermal printer mounted to the front of the camera. (If none of my friends are available, the camera uses Amazon's Mechanical Turk service, which lets me pay someone to write the description.)

The camera uses the *mrBBIO* Python module to read the shutter button and control the status LEDs. I used other Python modules for outputting descriptions as text on the printer and taking care of the networking that needs to happen in order to crowdsource the descriptions.

With all the features that the BeagleBone afforded me, and despite the learning curve, I think I chose the right technology to get the job done effectively. Creating a project like this with a microcontroller probably would have been a tough thing to do. ◪

⊞ To dig deeper into embedded Linux, check out elinux.org, which covers many different embedded Linux platforms. For specific help with the BeagleBone, try reaching out to the BeagleBone mailing list at groups.google.com/group/beaglebone or connect to the #beagle channel on the Freenode IRC network.

Matt Richardson is a contributing editor of MAKE and a Brooklyn-based technophile, maker of things, photographer, and video producer. His work can be found at mattrichardson.com.

Projects Editor Keith Hammond and his dog, Gage, prepare to hot-smoke some tasty tri-tips.

# The
# Nellie Bly
# Smoker

Written by
**William Gurstelle**

Photo by
**Gregory Hayes**

**Make a hot/cold food smoker from a 55-gallon steel drum.**

⚡ **TIME:** 1 WEEKEND ⚡ **COMPLEXITY:** MODERATE

Food prepared in a smoker is always a treat, so building a backyard smoker is a perfect project for those who love to combine making things with eating things.

This project is primarily an exercise in sheet metal work. You may need to purchase some tools and learn some new skills. Fortunately, the tools are relatively inexpensive and the skills not hard to learn. Plus, there's the benefit that, once obtained, both the tools and the skills will likely be useful for myriad future projects.

This electric smoker incorporates several useful features, including multiple doors and a large smoking area. The most interesting feature is the separate, movable firebox. By adjusting the distance between the firebox and the smoke chamber, the backyard charcutier can experiment with hot, warm, and cold smoking.

# Smoking Hot (and Cold)

The Nellie Bly Smoker is an electric smoker, and unlike most drum smokers it's got a traditional two-box config-uration. This design allows excellent temperature control.

Inside the **firebox** Ⓐ an **electric hot plate** Ⓑ heats wood chips in a **shallow pan** Ⓒ to generate smoke. A **louver** Ⓓ in the bottom controls airflow.

The food box or **smoke chamber** Ⓔ has 2 **sealed doors** Ⓕ for access, a **grill** Ⓖ to support food, and 4 **eyebolts** Ⓗ for hanging food. Two **thermometers** Ⓘ monitor the temperature inside.

A flexible, extensible **duct** Ⓙ carries smoke from the firebox to the food box. The temp-erature inside the food box is controlled by shortening or lengthening the smoke duct.

To help draw the smoke upward over the food, the food box is raised above the firebox by a **stand** Ⓚ, and fitted with a **chimney** Ⓛ.

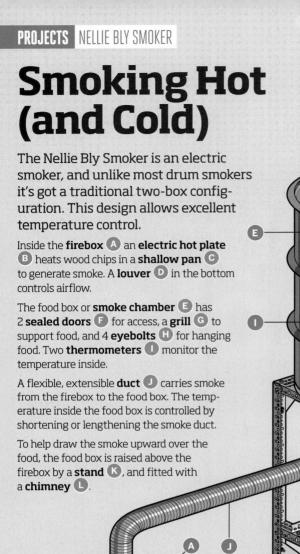

## Nellie's Drum

What a wonderful thing is the 55-gallon drum. Said to have been invented by the famous adventurer, reporter, and industrialist Nellie Bly, it's a great foundation for many maker projects and a cheap, utilitarian form of sheet metal.

Made from painted mild steel, standard 55gal drums are about 35½" tall and 24" in diameter, and while the thickness varies, most are made from about 18- or 19-gauge steel.

You'll find 2 main kinds of drums: *open head* and *tight head*. Open head drums, which are less expensive, are the kind you want for this project. They come with a removable top, held in place by a metal ring that's clamped with a large bolt.

In addition to the top and bottom lips, there are 2 raised ribs on the surface of the drum. These are called *chimes* in barrel-speak, and they're there to add strength to the cylinder.

Rob Nance

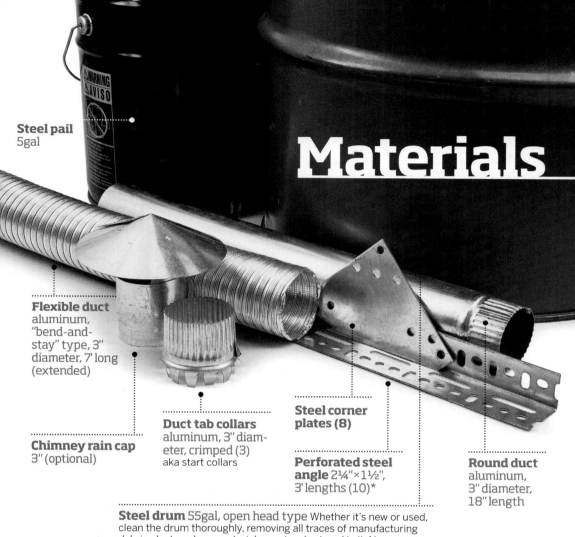

**Steel pail**
5gal

# Materials

**Flexible duct**
aluminum,
"bend-and-
stay" type, 3"
diameter, 7' long
(extended)

**Chimney rain cap**
3" (optional)

**Duct tab collars**
aluminum, 3" diam-
eter, crimped (3)
aka start collars

**Steel corner
plates (8)**

**Perforated steel
angle** 2¼"×1½",
3' lengths (10)*

**Round duct**
aluminum,
3" diameter,
18" length

**Steel drum** 55gal, open head type Whether it's new or used,
clean the drum thoroughly, removing all traces of manufacturing
debris, dust, and any materials previously stored in it. Never use
a barrel that's been used to store dangerous materials.

---

» **Hinges, 2" (6)**
» **Draw-pull latches, 3" (3)**
» **Silicone weatherstripping
  gasket, D-shaped, 12'**
» **Expanded metal, ⅛"
  thick, 2'×2'**
» **Steel rods, threaded,
  ¼-20, 18" long (2) with
  nuts (4) and washers (4)**
» **Eye bolts, ¼"×2" (4) with
  nuts (2) and washers (4)**
» **Sheet metal, 26 gauge,
  12"×24"**

» **Sheet metal, 22 to 26
  gauge, about 4"×8"**
» **Electric hot plate, 1,000W
  or more, adjustable,
  approx 9" diameter** We
  used Maxam #KTELSB.
» **Bolts, ⅜"×¾" (50) with
  nuts and lock washers**
» **Aluminized tape, 2" wide**
» **Meat thermometers (2)**
» **Steel pan, 10"diameter,
  shallow**
» **Machine screws or sheet
  metal screws (optional)**

# TOOLS

» **Jigsaw with metal cutting blades**
  You can also use a reciprocating saw,
  aka saber saw, but it's less accurate.
» **Drill and drill bits**
» **Center punch**
» **Socket wrench set**
» **Blind rivet tool with ⅛"
  or ³/₁₆" rivets**
» **Crayon or grease pen, light colored**
» **File, angle grinder, or rotary tool
  with grinding bit** e.g. Dremel
» **Bench vise**
» **Nibbler, electric or pneumatic
  (optional)**
» **Screwdriver (optional)**

*Wood or welded angle iron may be used for the stand in lieu
of perforated angle iron. But if you use wood, be aware that the
elevated smoke chamber temperatures could cause a fire.

Gregory Hayes

# Build your smoker.

## 1. MAKE THE DOORS

This diagram shows the modifications required to turn your 55gal drum into a working smoke chamber. In a nutshell, you need to make several openings in the barrel: the doors through which the food is inserted and removed, a smoke inlet hole, a smoke outlet hole, and several smaller holes for grill supports and thermometers near the doors.

**1a.** Using a light-colored grease pen or crayon, mark the doors as shown above, centered between the barrel chimes. Punch your hole marks before drilling, to center your drill bit and prevent it walking across the metal. In the corners of the door, drill holes large enough for the jigsaw blade.

Support the barrel so it stays in place, insert the jigsaw, and cut out the doors carefully — you'll use the cutouts as the doors. Grind all edges smooth with a file, rotary tool, or angle grinder.

**1b.** Use a nibbler or jigsaw to cut 1"-wide metal door facings from 26-gauge sheet metal, 1" longer and 1" wider than your doors. The doors close against these strips, keeping the smoke inside the chamber.

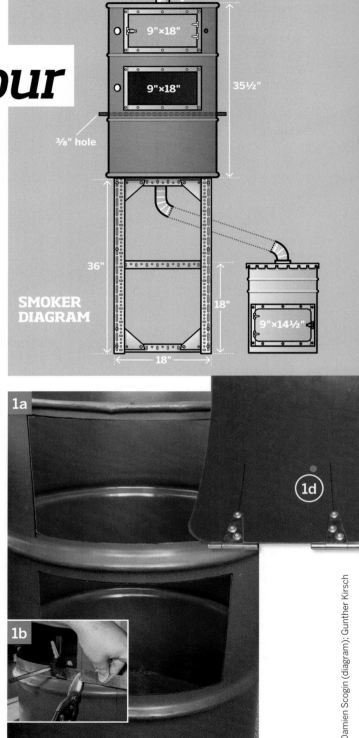

SMOKER DIAGRAM

18"

9"×18"

9"×18"

35½"

⅜" hole

36"

18"

18"

9"×14½"

**1c.** Place the facings on the barrel so they overlap the door opening by ½", and clamp them in place.

Drill ⅛" holes through the facings and barrel at 3" to 4" intervals, and then use ⅛" pop rivets to fasten the facings to the the barrel. You can also use ³⁄₁₆" rivets. If you've not used blind rivets before, it's easy and fun.

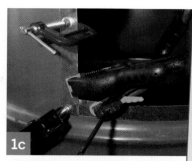

**1d.** Attach each door to the barrel with 2 hinges, using rivets, sheet metal screws, or short #8 machine screws. If you rivet the hinges to the doors, insert the rivets from the inside so they won't block the doors from closing.

**1e.** Attach the draw-pull latches to the barrel and door so that when the door is closed, the latch pulls the door securely into place.

**1f.** Once the doors are attached to your satisfaction, apply weatherstripping to the edges of the doors so smoke can't escape.

## 2. VENT THE SMOKER

**2a.** Cut two 3"-diameter holes, centered in the top and bottom flat surfaces of the barrel. Again, use a punch and drill to make a starter hole for the jigsaw.

Because jigsaws have a hard time with tight-radius cuts like this, you may want to use an air nibbler instead, to make quick work of the job.

**2b.** Remove the top from the barrel. Insert the tabbed collars into the top and bottom smoke holes and bend the tabs back so the collar stays securely in place. Seal with aluminized tape. Replace the top on the barrel.

# Pop Riveting
## A Useful Skill for Sheet Metal Work

A *blind rivet* or *pop rivet* gun has one fixed and one movable handle and an opening in the top that accommodates heads for different-sized rivets. Pop rivets are deformable tubes with a long pin (called the *mandrel*) through the middle. To permanently fasten 2 pieces of sheet metal:

**1.** Clamp the pieces to be fastened, and drill a snug hole through both (use a ⅛" bit for ⅛" rivets, etc.).

**2.** Insert the rivet in the hole, tube side first.

**3.** Insert the mandrel pin into the rivet gun and squeeze hard on the handles. The gun pulls the pin, which causes the tube on the rivet to deform, mushrooming outward and locking the 2 pieces of sheet metal. You'll hear a "pop" that means the pin has broken off and the 2 pieces are joined permanently.

William Gurstelle (pop riveting sidebar)

# 3. ADD THE GRILL, HANGERS, AND THERMOMETERS

**3a.** Drill ⅜" holes in the sides of the barrel as shown in the Smoker Diagram (page 98). Insert 18" threaded steel rods and secure with nuts. Cut a circular grill from expanded metal, sized to fit your barrel, and place it atop the rods.

**3b.** Drill four ⁵⁄₁₆" holes in the lid. Insert the eye bolts inside and fasten each with a nut and 2 washers.

**3c.** Drill holes near the doors and insert the thermometers such that they fit snugly.

# 4. MAKE THE CHIMNEY

Rivet the (optional) rain cap to one end of the 18" duct. Rivet the other end to the tab collar in the barrel lid, and seal with aluminized tape.

# 5. MAKE THE FIREBOX

**5a.** Test-fit your hot plate and pan in the bottom of the steel pail, then cut and mount a 9"×14½" door in the pail's side, as in Step 1.

**5b.** Lay out 2 triangular vent holes in the bottom of the pail, and a 3" smoke outlet hole centered in the lid. Use the punch, drill, and jigsaw to cut out the holes.

Insert the remaining tabbed collar into the 3" round hole in the top of the firebox, bend the tabs to secure it, and seal with aluminized tape.

**5c.** Cut a louver from thin sheet metal large enough to cover both vent holes. Attach to the bottom of the firebox with a rivet in the center.

**3a**

**TIP** These eye bolts are useful for hanging large fish and fowl in the smoke chamber.

**3b**

**5b**

**5c**

Damien Scogin (diagram)

# 6. BUILD THE STAND

The smoke chamber must be positioned higher than the firebox. I built a sturdy stand from perforated steel angle and bolts, but you could weld angle iron instead, or improvise your own frame.

The Smoker Diagram shows how to lay out and assemble the perforated steel angle. Use angle plates at the corners to add strength and rigidity to the structure.

Note the position of the cross-members about midway between top and bottom. These support the firebox during hot smoking.

# 7. CONNECT THE SMOKE DUCT

**7a. Hot smoking setup.** Place the firebox on the stand directly below the smoke chamber's inlet hole. Connect the firebox outlet collar to the smoke chamber inlet collar using a short piece of 3" duct. Seal with aluminized tape.

**7b. Cold smoking setup.** Position the firebox about 6' away from the smoke chamber and extend the 3"-diameter bend-and-stay duct to its full length. Attach the duct to the smoke chamber inlet collar and the firebox outlet collar. Seal with aluminized tape.

➕ Want to get started? Check out our recipe on how to smoke fish at makeprojects.com/v/32.

7a

7b

William Gurstelle is a contributing editor of MAKE. Visit williamgurstelle.com for more information on this and other maker-friendly projects.

⚑ **TEST BUILDER:** Daniel Spangler, MAKE Labs

William Gurstelle (7b)

# Smoke 'Em If You Got 'Em

## Hot Versus Cold

There are 2 main categories of smoking: hot and cold. In hot smoking, the smoke heats the food to about 126°F–176°F. Some hot-smoked foods, such as fish, may cook fully in the smoker. Other foods, such as red meat, should be pre-cooked. Consult a cookbook for details on safely preparing meats.

Since cold smoking (68°F–86°F) does not cook food, cold-smoked foods must be cured or cooked before being eaten.

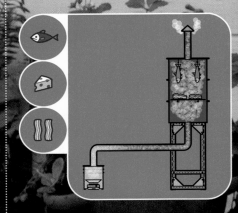

## So why cold-smoke at all?

Cold-smoked meats and fish are moister and often more flavorful. The choice of hot or cold smoking depends on the item being smoked, how the chef wants the food to taste, and the equipment and time available.

**Good woods** to use for smoking are hardwoods such as hickory, beech, alder, mesquite, and fruit and nut woods like apple and pecan. **Don't use** softwoods such as pine and fir because their resins produce undesirable chemicals in the smoke.

## Fuel and Temperature Control

Mound wood chips or sawdust in a shallow steel pan about 10" in diameter, so they make a cone-shaped pile. Place the pan on the electric hotplate. Adjust the heat on the hotplate so the chips smolder but don't burn with a flame. When the chips are used up and the smoke becomes thin, add more chips.

The temperature inside the smoke chamber is the critical variable in successful smoke cooking. This is controlled by the length of the **duct** between the firebox and smoke chamber, the **heat** of the firebox, and the ambient **air temperature**. Choose cool days for cold smoking and warm days for hot smoking. Try adding a little **charcoal** to the pan to sustain warmer temperatures.

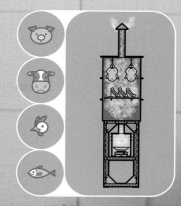

**Adjust the smoke levels** by opening or closing the louver on the bottom of the firebox to let in more or less air. This will speed up or slow down smoke production.

## Why Smoke?

Exposing foods such as meat, cheese, and fish to wood smoke is desirable for a few reasons. Wood smoke adds wonderful flavors, from sweet caramel notes to spicy, smoky, and even vanilla aromas. Smoking also dehydrates foods to some degree, changing the texture and making them saltier and savorier.

Finally, wood smoke contains antioxidant and antimicrobial compounds that slow the rates at which fats turn rancid and bacteria multiply, so smoking helps preserve foods.

**Consult** one of the many excellent cookbooks that address safe food preparation prior to smoking. It's imperative that you use the correct techniques to prepare your smoked food.

Gregory Hayes

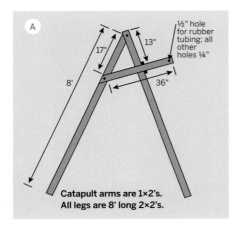

A

17"    13"

8'    36"

½" hole for rubber tubing; all other holes ¼"

Catapult arms are 1×2's.
All legs are 8' long 2×2's.

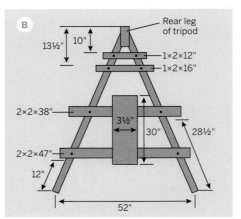

B

Rear leg of tripod

13½"    10"

1×2×12"
1×2×16"

2×2×38"

3½"

30"    28½"

2×2×47"

12"

52"

# Catapult Glider Launcher

## Fling your Rocket Glider or other toy aircraft 150 feet into the sky!

Written and photographed by **Rick Schertle**

For many years, the U.S. military has enjoyed playing with "toys." Today that includes unmanned drone aircraft, but in the past, it was toy balsa wood airplanes. Thousands of folding-wing balsa gliders were shot out of the sky in the early 1940s for World War II artillery practice. High in the air, these toy airplanes had the scale effect of a full-sized plane. Developed and patented in 1939 by Jim Walker, founder of the American Junior Aircraft Company, the folding-wing Army Interceptor glider bore the brunt of the action.

My folding-wing Rocket Glider, based on the Interceptor, was featured as a how-to project in MAKE Volume 31 and is available as a kit from Maker Shed. While the Interceptor originally used a handheld rubber-band catapult, the military designed a launcher to hurl the gliders nearly 300 feet high. In 2004, Frank Macy designed the first reproduction of this launcher, on which this project is based.

With nearly 20 pounds of pulling force, this simple catapult remarkably flings the tiny glider (weighing less than an ounce) 150 feet into the air. Stand clear, pull the rope trigger, and watch a piece of history rocket into the sky!

## 1. Build the tripod catapult stand.

**1a.** Cut the lumber to the following lengths:
» 2×2: 8' (3), 38" (1), and 47" (1)
» 1×2: 36" (2), 12" (1), 16" (1), 17" (1), and 3" (1)
» 1×4: 30" (1)

# Materials

- **Lumber: 2×2×8' (4), 1×2×8' (2), 1×4×30" (1)** Cheap pine is fine; I used clear fir for a nicer look.
- **Bolts, ¼": 4" (5), 3½" (4), 5½" (1)**
- **Nuts, ¼" (8)**
- **Wing nuts, ¼" (2)**
- **Washers, ¼" (20)**
- **Wood screws: ¾" (5), 2" (4)**
- **Screw eyes, 1⅝" (4)**
- **Chain, light duty, 33"**
- **Wood glue**
- **Hinge, T style, 4"**
- **Aluminum C-channel, ½"×½"×¹⁄₁₆" thick, 26" length**
- **Surgical tubing, ½" OD, 6' length**
- **Steel rod, ⅛" diameter, 9½" length**
- **Wire or cable, 14 gauge, stranded, coated**
- **Rubber band, 3½", heavy duty** or you can double a 7" band
- **Rope, light duty, 6'** needs to slip smoothly through screw eyes
- **Wire clothes hanger**
- **Folding-wing glider** for launching. Make your own at makeprojects.com/project/f/1934, or get our Rocket Glider kit, item #MKRS2 at Maker Shed (makershed.com).

## TOOLS

- **Miter saw** power or hand
- **Drill and drill bits:** ³⁄₃₂", ⅛", ¼", ½"
- **Hacksaw**
- **Screwdriver and wrench or socket set**
- **Wire cutters/strippers**
- **Scissors or utility knife**

Build the glider in
**MAKE VOLUME 31**

*Or get the kit from*
**MAKER SHED #MKRS2**

↗ **TIME:** A WEEKEND    ↗ **COMPLEXITY:** EASY

Gregory Hayes

**1b.** Assemble the tripod stand, following the assembly drawings (**Figures A and B**). Cut the top inside corners of the 2 front legs at 15° so they'll meet the rear leg flush when they're splayed out (**Figures C and D**).

Drill ¼" holes where indicated, then attach the pieces with ¼" bolts, washers, and nuts, except the 30" launch platform (see Step 2).

Use wing nuts at the top of the tripod (with a 4" bolt) and where the catapult arms attach to the rear leg (5½" bolt). This way, the catapult arms can be loosened and swung back so the launcher is more portable.

**1c.** Attach one screw eye 3' up on the rear leg, and another on the upper rail that supports the launch platform. Measure 33" of chain and attach it to both screw eyes (**Figure E**).

Drill ½" holes horizontally through the free ends of the catapult arms.

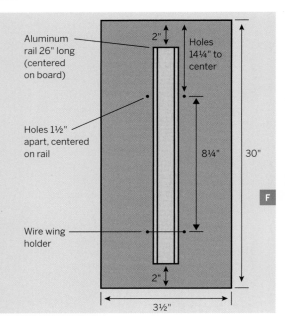

Aluminum rail 26" long (centered on board)

2"

Holes 14¼" to center

Holes 1½" apart, centered on rail

8¼"

30"

Wire wing holder

2"

3½"

F

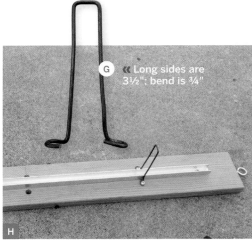

G « Long sides are 3½"; bend is ¾"

H

## 2. Build the launch platform.

**2a.** Cut the aluminum C-channel to 26" using a hacksaw. Following **Figure F**, place the channel, open side up, in the center of the 1×4 board, 2" from the top. Drill pilot holes through the channel and attach it with ¾" wood screws.

**2b.** Measure down 14¼" from the top of the 1×4, then drill two ¼" holes 1½" apart (on center), one on each side of the channel.

**2c.** To install the wing holder, cut a 10" piece of wire from a clothes hanger and bend it into the shape shown in **Figure G**. Then screw it into the launch platform 8¼" below the ¼" holes you made. Placement is important; it must hold the Rocket Glider's wings back, but

not impede it from leaving the launcher.

If you'll be launching some other craft, don't install the wing holder yet; wait until you build the trigger mechanism in the next step so you can get your measurements right.

**2d.** Attach a screw eye centered in the bottom edge of the launch platform (**Figure H**); this will guide the trigger cord.

**2e.** Finally, mount the launch platform centered on the lower horizontal rails on the tripod stand, using wood screws.

## 3. Make the trigger.

**3a.** Glue and screw the 3" piece of 1×2 horizontally to the top of the 17" piece of 1×2 (**Figure I**). Drill two ³⁄₃₂" or ⅛" holes in the horizontal piece, 1½" apart, so the ⅛" rod will fit snugly.

**3b.** Bend the ⅛" rod into a U shape with 4" legs, 1½" wide at the base. Carefully hammer the U-shaped rod all the way into the 2 holes so they stick out the other side

I

J

# IMPORTANT

Make sure the trigger rods are the same length, so they'll release the glider at the same time.

about 3" (**Figure J**). These are the trigger rods.

**3c.** Drill a ½" hole through the bottom of the trigger board and tie the launch cord through it.

## 4. Put it all together.

**4a.** Feed the trigger rods through the ¼" holes in the launch platform. Attach 2 wood screws into the sides of the launch platform and stretch a rubber band between them to hold the trigger board in place. The 2 pins should move easily in and out of the holes in the launch platform. Bend them if you need to.

When you're satisfied with the trigger action, attach the T-hinge to the trigger board and the bottom rail (**Figure K**).

**4b.** Feed the trigger cord through the screw eye in the bottom of the launch platform.

**4c.** Cut the surgical tubing into two 3' lengths and tie a big knot in one end of each. Feed these through the ½" holes in the ends of the catapult arms.

**4d.** Cut a 24" length of the heavy coated wire and tie a knot in each end. Tie the wire into the free ends of each length of surgical tubing (**Figure L**). You're done. ◪

---

Rick Schertle (schertle@yahoo.com) teaches middle school in San Jose, Calif., and designed the Compressed Air Rockets for MAKE Volume 15 and the Rocket Glider for MAKE Volume 31. With his wife and kids, he loves all things that fly.

## Launching Tips

Set up your launcher on a good-sized field with little or no wind. With the catapult arms in launch position, tighten the wing nuts.

Pull the surgical tubing downward and hook the catapult wire onto the trigger rods (Figure L). The tubing should be pointing straight up vertically.

Now practice triggering the catapult without the glider. With your head well clear, place one foot on the catapult cross-rail, and *slowly* pull the rope back, triggering the catapult. Don't jerk it, or the trigger can bounce back and damage your glider.

This is an extremely powerful catapult to launch a glider that weighs barely 1oz. There is room for disaster. I suggest practice-launching something else before you move on to the glider.

Now for the true test. With the wings folded back, place the glider in the C-channel, and hook the launch notch in the bottom of the glider onto the catapult wire. The folded wings should slip freely into the wire wing holder (**Figure M**).

Just like the U.S. Army did 70 years ago with a similar launcher, pull the rope *slowly* … and watch your glider zip skyward.

## Mods

This catapult launcher is a rough replica of the one used by the U.S. military, so while it folds, it's still not very portable. I challenge MAKE readers to come up with more compact and clever designs and share them at make projects.com/project/g/2563. The possibilities are endless!

CAUTION Once the launcher is active, keep your head away from the catapult!

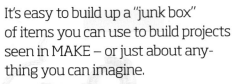

# Computer Printer Salvage

## PC Load Letter?!
## Over 200 useful parts for free!

By *Thomas J. Arey*

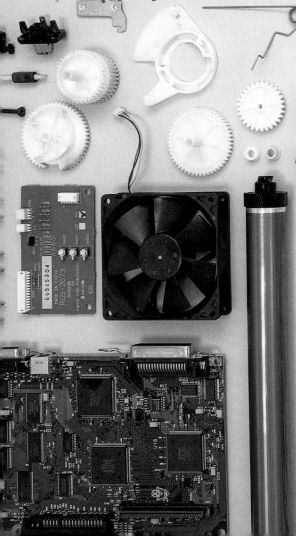

It's easy to build up a "junk box" of items you can use to build projects seen in MAKE – or just about anything you can imagine.

Many of my articles for MAKE (makezine. com/go/arey) take advantage of found components, often picked out of trash bins. Just because an electronic device has failed at its original task doesn't mean it can't perform other tasks. Castoffs can be recovered and the parts repurposed in countless ways.

Recently, my trash-picking adventures turned up a discarded laser printer. I set about finding what wonders were waiting beneath the plastic covers.

### JACKPOT OF PARTS

My first discovery was the main circuit board. Once I stripped the heat shields off, I found over 50 nonproprietary electronic parts, including **capacitors**, **resistors**, **voltage regulators**, **transistors**, **transformers**, **coils**, and **integrated circuits**. Jackpot! A couple of boards like this, and you're on your way to building a backup supply of parts for future projects. A second, smaller PC board also yielded numerous useful components.

**WARNING** Capacitors can hold a charge for months after they've been disconnected. Always bleed off the charge to ground, to prevent any shock hazard.

There are 2 strategies for keeping these components in stock: take the time to desolder the parts and store them separately, or leave them on the boards and inventory them on a sheet of paper attached to the board for future reference.

Digging deeper into the printer, I uncovered the paper transport mechanism. Gears, gears, and more gears! I harvested 2 **gear mechanisms** for turning rotating movement into linear movement. Anyone interested in robotics projects would find these useful.

The rest of the teardown revealed the **laser tube** and **lenses**, **various rollers**, **wires**, **springs**, and **hardware**, **micro switches**, **relays**, and of course a **motor**. I scored more than 200 free, useful parts for future use.

And don't forget the plastic or metal case and chassis parts. I keep a couple boxes of these allegedly useless scraps to cut and shape when developing design ideas. ◪

➕ For more on salvage, inventory, and storage, go to makezine.com/32/printersalvage.

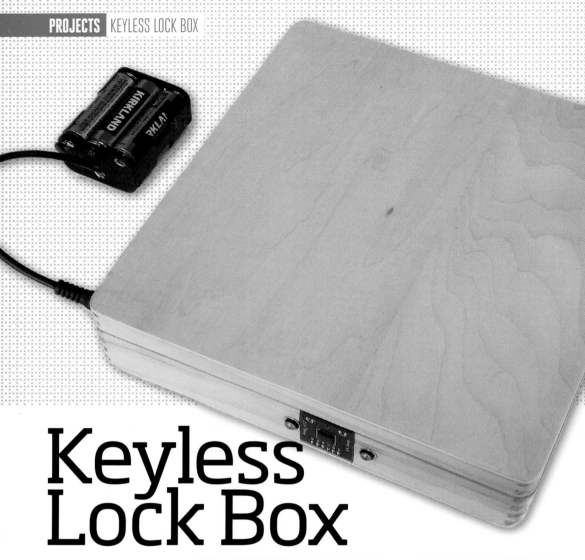

# Keyless Lock Box

**A wave of your finger opens this magic treasure chest.**

Written, photographed, and illustrated by
*Gordon McComb*

⚡ **TIME:** 4 HOURS  ⚡ **COMPLEXITY:** MODERATE

That piggy bank is looking mighty low-tech these days, and you have to bash it open to get your cash. One piggy, one withdrawal. And then there are those cheesy, tin "Wild West" lock boxes with the red combination dial. Not only do they lack wow factor, your granny could crack one in 30 seconds.

But stick an Arduino in a wooden box, along with a finger-operated sensor and small motor, and you've made a 21st-century treasure chest that's suitable for a daily diary, petty cash, or even those special Rice Krispies recipes that your snoopy neighbor wants to steal.

There's no key in this keyless electronic combination lock box; you just move your finger across a small optical window, and it's

# How It Works

» The keyless lock box uses a **unique optical finger navigation (OFN) sensor** as a combination decoder. The OFN sensor works much like an optical mouse, except it's intended to be used in direct contact with your finger. They are used in handheld devices where a trackpad would be too large, but because they are more expensive than trackballs, they're not common in consumer products.

» Movement across the small surface of the sensor is converted to **X and Y distance measurements** – up, down, left, and right. Sequences of these movements make up the combination of the lock.

» For this project I'm using the **Parallax OFN module**, which puts a commercial OFN sensor on a breakout board that provides connectors for power (3.3V to 5V), ground, and 6 signal lines. The OFN module uses 2-wire I2C to communicate with a microcontroller, and has additional I/O pins for such things as the momentary pushbutton switch that engages when you push the optical sensor down.

» The locking mechanism uses a standard-size **R/C servomotor** that's glued into the bottom of the box. To lock the box, the turning servo engages a metal arm attached inside the box's lid. Turning the other way, it frees the bar, letting you open the lid.

» An **Arduino microcontroller** works as the main brain of the lock box, handling all the communications with the OFN module, controlling the servo, and even making musical tones on a small piezo speaker.

» For my box, I used a plain 8" square **cigar box** from a craft store – no need to smoke a bunch of stogies. The wood is unfinished; stain or paint to suit. You don't get Fort Knox with these boxes, but they'll keep out the casual thief.

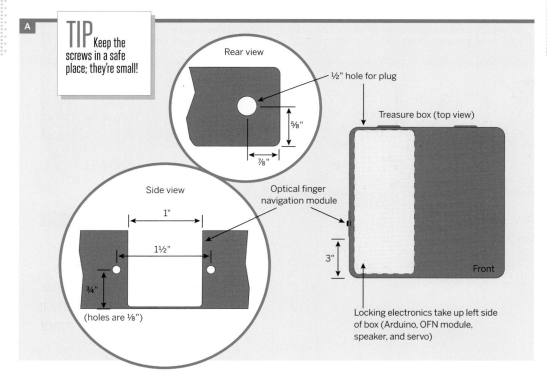

**A**

**TIP** Keep the screws in a safe place; they're small!

Rear view

½" hole for plug

⅝"

⅞"

Treasure box (top view)

Side view

1"

1½"

¾"

(holes are ⅛")

Optical finger navigation module

3"

Front

Locking electronics take up left side of box (Arduino, OFN module, speaker, and servo)

"open sesame." The combination to the lock is a secret movement pattern only you know.

## 1. Prepare the box.

To make cutting and drilling easier, detach the box lid by removing the top screws in the 2 hinges on the back. Use a hacksaw or razor saw to cut out a small chunk on the left side of the box for the OFN module, just wide enough to fit (**Figure A**). Gently pry loose the knockout piece, being careful not to crack the bottom of the box.

# Materials

» **Cigar box, or similar wooden box**
Mine measured 8"×8"×1¾"
» **Arduino Uno microcontroller board**
» **Arduino IDE** Version 1.0 or higher, arduino.cc
» **Optical finger navigation (OFN) module**
Parallax #27903, parallax.com
» **Power source, 9V DC, 1,000mA, with 2.1mm center-positive plug** Plug-in power adapter or battery holder, either 6×AA or 9V. If you prefer to use rechargeable NiCd or NiMH batteries, you need an 8×AA battery holder.
» **Piezo speaker, 5V** I used BG Micro #AUD1112.
» **Parallax/Futaba servomotor, 6V, standard R/C size** Parallax #900-00005
» **Capacitor, tantalum electrolytic, 47µF, 15V or higher**
» **Resistors, 2.2kΩ, ¼ watt, 5% tolerance (2)**
» **Circuit board, 780-hole component layout**
RadioShack #276-168, radioshack.com
» **Wire jumpers, 6": male/male (9), male/female (3)** I like to use stranded wires; they're more flexible and less likely to break after repeated bending.
» **Header, double-length, 3-pin**
» **Header shells, 1×1 (16)** aka crimp connector housings
» **Cable, 3-wire, male/female, 14"** for hobby servos
» **Plywood, aircraft-grade, ¼"×3½"×2"**

# TOOLS

» **Soldering pencil, 25–30 watt, fine tip**
» **Solder**
» **Flush cutters** for wire
» **Needlenose pliers**
» **Drill with drill bits: 1", ¼"**
» **Step drill bit, including ½" size**
e.g. 6" to ½"
» **Hacksaw or hobby razor saw**
» **Screwdriver, Phillips head, #1**
» **Hot glue gun and glue** low-temp glue recommended
» **Florist tape (putty), ½" square**

or expanded PVC
» **Brass strip, 3⅝"×½"×0.064"**
» **Machine screws, 4-40: ⁷⁄₁₆" flat head (4), ⅝" pan head (2), and ¾" pan head (1)**
» **Nuts, 4-40: steel (2), nylon (2), nylon-insert locking (1)**
» **Nuts, #4, ⅛", and washers, #4 (2 each)**
» **Hookup wire, 22 gauge, solid, insulated**
» **Heat-shrink tubing, ⅛"**
» **Double-sided foam tape**

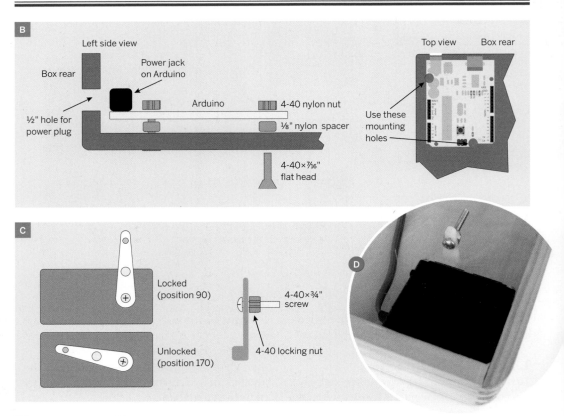

**B**

Left side view

Box rear

Power jack on Arduino

½" hole for power plug

Arduino

4-40 nylon nut

⅛" nylon spacer

4-40×⁷⁄₁₆" flat head

Top view    Box rear

Use these mounting holes

**C**

Locked (position 90)

Unlocked (position 170)

4-40×¾" screw

4-40 locking nut

**D**

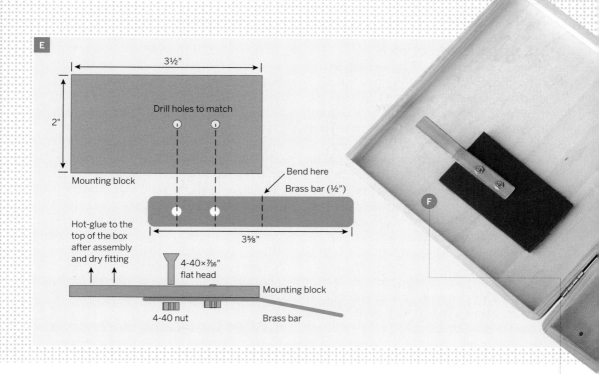

**E**

3½"

2"

Drill holes to match

Mounting block

Bend here

Brass bar (½")

Hot-glue to the top of the box after assembly and dry fitting

3⅝"

4-40×7⁄16" flat head

4-40 nut

Mounting block

Brass bar

**F**

Drill ⅛" mounting holes for the OFN module. To prevent the bit from pushing through and splintering the wood, press or clamp a piece of wood behind the hole while drilling. Measure and mark the center of a ½" hole in the rear of the box, positioned so you can insert a barrel plug through the hole and into the Arduino's power jack. First drill a pilot hole with a ⅛" bit. Follow that with a ¼" bit, and finish with a step bit, stopping at the ½" mark.

Referring to **Figure B**, temporarily position the Arduino in the rear left of the box, on top of ⅛" nylon spacers. Make sure the Arduino power jack is in line with the rear hole you just drilled. Use a small nail or sharp pencil to mark the location for the 2 mounting holes indicated. Remove the Arduino and set aside.

## 2. Make the locking mechanism.

Alignment of the locking parts is critical, so take this step slowly. Use a hacksaw to carefully remove the mounting flanges on both sides of the servo. Thread a 4-40×¾" machine screw near the end of the included single-arm servo horn and use a 4-40 self-locking (nylon insert) nut to tightly secure the screw in place (**Figure C**).

Attach the horn to the servo and slowly rotate the motor shaft to the center of movement. You'll make fine-tune adjustments later, so leave off the screw that holds the horn to the servo shaft. Use a hot glue gun to mount the servo in the front left corner of the box (**Figure D**).

To avoid unsightly screws on the top of the box, fasten the locking bar to a mounting block made of ¼" aircraft-grade plywood or PVC inside the lid. The large surface area of the flat block will enable a strong glue joint.

Follow **Figure E** to cut and drill a length of ½"-wide × 0.064"-thick brass strip, then cut the block to size, and drill holes matching those in the strip. Fasten the brass strip and block together with 4-40 flat-head machine screws. Bend down the front of the strip at a 15°–20° angle. You can adjust the angle later to achieve better locking action.

Stick a single ½" square piece of florist tape (putty) to the top of the block, and using your best guess, line up the locking bar so it engages with the screw attached to the servo horn. The putty keeps the mounting block in place until you can test the best placement on the lid. Reattach the lid to its hinges (**Figure F**).

## 3. Calibrate the servo.

Wiring makes or breaks a project. I used pre-crimped male/male and male/female jumpers, 1×1 header shells, and double-long breakaway header pins for robust and easily pluggable connections. You could theoretically use the pin sides of 3 male/female jumpers to connect to the servo extension cable, but the double-long headers make a stronger friction-fit connection. Snap shells onto the jumpers to make 9 male/male jumpers with shells at one end, and 3 male/female jumpers with shells on both ends (**Figure G**).

I needed to add a bypass capacitor to reduce electrical noise from the servo that caused the Arduino to keep resetting itself. Solder a 47µF tantalum capacitor between the middle and one side pin of a 3-pin double-long header (**Figure H**). Be *absolutely* sure that the + lead of the capacitor is connected to the center pin! These will connect to the servo's voltage (V+) and ground (Gnd) pins.

To calibrate the servo so that it points in the desired direction, download and unpack the project code from makeprojects. com/v/32, then verify (compile) and upload the *ServoCalibrate* sketch to your Arduino. Visit the same link for instructions on how to

upload programs to an Arduino.

Remove the horn from the shaft and plug in the 3-pin connector you just made, orienting the bypass cap on the black or brown wire (ground) side. Use 3 male/female jumpers to connect the servo and Arduino's 5V and ground together, and the servo's control (white or yellow) to Arduino digital pin 8 (**Figure I**).

Briefly depress the reset switch on the Arduino. When the *ServoCalibrate* sketch restarts, the servo first moves to its extremes, then centers itself at its midpoint and stops. After it's done, unplug the Arduino from the PC, then reattach the horn to the servo so that it points straight up, and secure it in position with the small included screw.

## 4. Align the locking mechanism.

To align the servo horn with the brass strip, reconnect the Arduino to your PC and upload the *ServoLock* sketch. Thread the servo cable through the opening in the side of the box, manually move the servo arm toward the front of the box, and close the lid. Click the reset button to run the sketch and listen for when the servo stops moving. Then test the lock by trying to lift the lid.

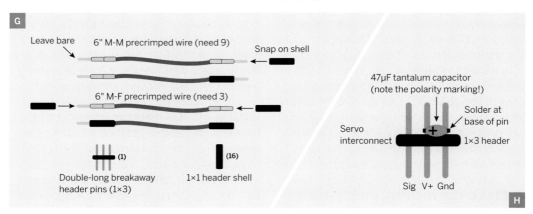

G

Leave bare

6" M-M precrimped wire (need 9)    Snap on shell

6" M-F precrimped wire (need 3)

(1)    (16)

Double-long breakaway    1×1 header shell
header pins (1×3)

47µF tantalum capacitor
(note the polarity marking!)

Solder at
base of pin

Servo
interconnect    1×3 header

Sig  V+  Gnd

H

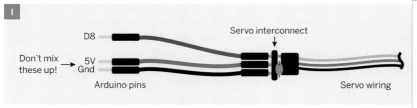

I

D8

Don't mix    5V
these up!    Gnd

Arduino pins

Servo interconnect

Servo wiring

**CAUTION!**
Warning! Whoa, Nellie!
Do not cross the
polarity of the wires
or your servo may be
permanently damaged.

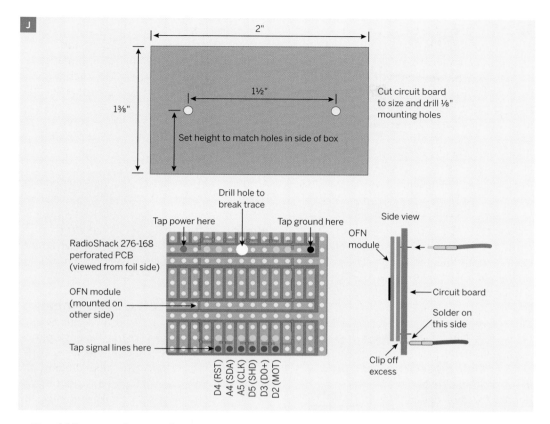

J

2"

1½"

1⅜"

Set height to match holes in side of box

Cut circuit board
to size and drill ⅛"
mounting holes

Drill hole to
break trace

Tap power here          Tap ground here

Side view

RadioShack 276-168
perforated PCB
(viewed from foil side)

OFN
module

OFN module
(mounted on
other side)

Circuit board

Solder on
this side

Tap signal lines here

Clip off
excess

D4 (RST)
A4 (SDA)
A5 (CLK)
D5 (SHD)
D3 (D0+)
D2 (MOT)

Should the servo horn and screw not properly engage over the brass strip — it makes clunking or scraping noises — pop the mounting block off the lid, reposition it, and try again. When you get the position just right, unplug the servo and mark around the block with a pencil for gluing later.

## 5. Install the OFN board, speaker.
The OFN module doesn't have any direct means for mounting, so you need to solder it to a small circuit board that you can then attach to the side of the lock box using 4-40 machine screws and nuts. Cut the board to size and drill 2 holes for mounting (**Figure J**).

I used a board with component layout, in which long horizontal bus traces alternate with 3-hole vertical segments. This layout makes it easier for making wire connections on the backside of the board with the copper foil, which is necessary because the OFN module will take up all the space on the front.

Following Figure J, drill a ⅛" hole in the

middle of the long trace at the top, to break the connection. This will separate the voltage and ground connections to the OFN module, which is very important. Then solder the OFN module to the board from the top, non-foil side (the leads will poke through to the foil side). Orient it with the LED on the drilled trace side, along with voltage and ground connections, and the 6 signal lines along the opposite side.

Cut eight 1" lengths of ⅛"-diameter heat-shrink tubing, and slip one each over 8 of your prepared male/female jumpers. Solder the bare pin end of each jumper to its correspond-ing connecting point on the foil side of the board 6 signal lines in a row, and the power and ground on the other side, as seen in **Figure K**, page 116. The pin will poke through; cut it flush with the front of the board after soldering. After all jumpers are soldered, push the heat-shrink tubing over the metal and apply heat to shrink. Mount the OFN module and board to the side of the box using 4-40 fasteners (**Figures L and M**).

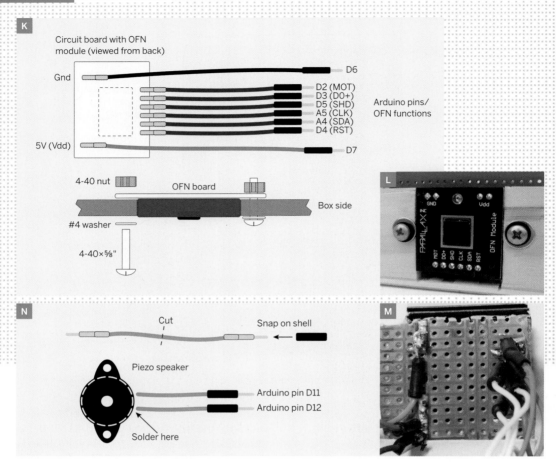

Circuit board with OFN module (viewed from back)

Gnd

D6

D2 (MOT)
D3 (DO+)
D5 (SHD)
A5 (CLK)
A4 (SDA)
D4 (RST)

Arduino pins/
OFN functions

5V (Vdd)

D7

4-40 nut

OFN board

Box side

#4 washer

4-40×⅝"

Cut

Snap on shell

Piezo speaker

Arduino pin D11
Arduino pin D12

Solder here

Plug in the 8 jumpers from the OFN into the Arduino digital pins D2–D7 and analog pins A4–A5 as shown in Figure K. Be absolutely sure not to swap pins D6 and D7, or you'll reverse the power to the OFN and possibly damage it.

Cut the remaining male/male jumper in half. Strip ⅛" of insulation from the cut ends, and solder them to the terminals of the piezo speaker. Use double-sided foam tape to mount the speaker to the bottom of the box, and attach the 2 jumpers to Arduino pins D11 and D12 (**Figure N**). The innards of your lock box should look like **Figure O**.

## 6. Program it.

Reconnect your Arduino and then verify and upload the *LockBox* sketch. It includes several files in the same folder; make sure they remain in the folder or the sketch won't work.

*You need version 1.0 (or later) of the Arduino IDE, or the sketch will not compile.*

For now, keep the lid open so you can see what happens when the box locks and unlocks. Keep the Arduino connected to the computer and open the Serial Monitor window (Tools menu) so you can review the debugging messages.

On Arduino startup or reset, the servo should go into lock (straight up) position, and you should see "Device ready" in the Serial Monitor window. If instead it reads, "Device not found," it means the Arduino can't find the OFN module, and you should recheck the wiring.

Swipe your finger across the sensor of the OFN module sensor, then press it down. You'll hear a series of tones indicating a bad combination entry. Move your finger across the sensor again, this time using the pre-stored combination: *Left-Right-Left.* Click down the

# Lock Box Power!

The Keyless Lock Box is designed to use external power for unlocking, but it can sit in a locked state for indefinite periods without any power. To power the lock and open it, insert the plug into the hole in the back, enter the combination, then lift the lid.

The box will automatically relock after 7 seconds. If the lid is still open, you can relock the box by manually pushing the servo arm toward the front of the box, pressing the Arduino's reset button, and closing the lid. When the sketch restarts, it automatically moves the servo to its lock position. You may then remove the power plug.

**If you forget the combination or can't get it to work, here's a "secret" override procedure that will briefly unlock the box:**

Unplug power to the Arduino. Press and hold down the OFN sensor while you plug the Arduino back in. Quickly open the lid when you hear the servo move to its unlock position. You'll have only 1.5 seconds before the servo relocks the box.

You can remove or comment out the override code in *LockBox.ino* if you don't want this behavior.

sensor, and this time the lock opens. After 7 seconds the lock automatically closes.

See the comments in the *LockBox* sketch on where to change the combination. It can be any length sequence of swipes up, down, left, and right. Once the lock is reprogrammed, unplug the Arduino from your computer. Reattach the Arduino to the bottom of the lock box. Now's a good time to glue the locking bar mounting plate permanently to the lid. Hot glue or wood glue works best.

After the glue sets, stash your best secrets in your new keyless combination box. Whenever you need access, let your fingers unlock the mysteries within.

Keep the box looking good by applying paint or finish to the bare wood, starting with a base coat. For a cleaner look, remove the clasp in front of the cigar box, and fill in the holes with wood putty. Sand for a smooth finish. Or replace the clasp with something fancier. ◪

Gordon McComb has been building robots since the 1970s and wrote the bestselling *Robot Builder's Bonanza*. You can read his plans to take over the world with an army of mind-controlled automatons, along with other musings, at robotoid.com.

PROJECTS

**A hardware solution to help you when a synonym for "awesome" doesn't come to mind immediately.**

Written and photographed by *Matt Richardson*

↗ **TIME:** A DAY    ↗ **COMPLEXITY:** MEDIUM

Ever since I started writing for MAKE, I've kept an eye on all the awesome websites out there for awesome makers and the awesome projects that are posted every day. Luckily for me, there's no shortage of awesome work to write about. My only difficulty was I needed more words to describe how awesome this stuff is.

To fix this problem, I created the Awesome Button, my own custom USB input device that keys in a random synonym for awesome, on demand. With the Awesome Button, when I'm writing about a project that I like a lot and I get stuck on how to describe it, I hit the big red button on my desk and it takes care of the adjective. Now instead of awesome this and awesome that, I'm writing about incredible robots, fantastic camera hacks, and cool electronics projects.

# The
# AWESOME
# Button

**Switch, momentary** any style you like. I used a big red one, item #9181 from SparkFun (sparkfun.com).

**Rubber feet** to mount on bottom of enclosure

**Enclosure** appropriately sized for your button. I used item #270-1807 from RadioShack (radioshack.com). You could also use a spare cardboard box, or laser-cut a custom enclosure (see page 53).

# Materials

**Wire, insulated, 22 gauge**

**USB cable, Standard-A to Mini-B** You probably have a few spares lying around.

**Heat-shrink tubing** in various sizes. You could use electrical tape instead.

**Mounting tape, double-sided**

**Teensy USB Development Board, Version 2.0** $16 from PJRC (pjrc.com)

# Mighty Teensy

At the core of the Awesome Button is the Teensy USB Development Board, made by PJRC. This tiny USB microcontroller can be programmed to act like a USB Human Interface Device, such as a keyboard, mouse, or joystick. Since we want the Awesome Button to type words in for us, we'll set it up as a keyboard.

The main advantage of using this method is that the Awesome Button will work with any USB-equipped computer and any application that takes text input. Whether you're writing an email, working in a word processor, or chatting on IRC, the Awesome Button can key in those hard-to-think-of synonyms.

# Build your button.

## 1. INSTALL THE BUTTON

**1a.** Measure to find the center of your enclosure and drill a hole for your button. For the SparkFun button, the diameter should be 1".

**1b.** Remove the snap-action switch and the LED assembly from the bottom of the button.

**1c.** Place the button through the hole, screw the ring down, and replace the snap-action switch and LED assembly.

## 2. WIRE THE SWITCH

**2a.** Solder a 6" length of wire to the terminal marked "NO," which means the circuit between this terminal and the common terminal is "normally open." When we press the button, it will close the connection between the common and NO terminals.

**2b.** Solder another 6" wire to the terminal marked "Common."

## 3. CONNECT THE SWITCH TO THE TEENSY

**3a.** Solder one wire from the button to a ground pin on the Teensy.

**3b.** Solder the other wire from the button to the pin marked B0.

**3c.** Attach the Teensy to the enclosure using a small piece of mounting tape.

1a

1c

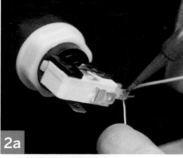

2a

NOTE There's no pull-up or pull-down resistor for this digital switch circuit because the code will activate the internal pull-up feature of the ATmega32U4 chip on the Teensy USB board.

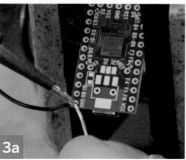

3a

3c

# 4. MOD THE ENCLOSURE

**4a.** Drill a hole just wide enough for your USB cable to fit through.

For a cleaner look (optional), make this hole the width of the cable itself, not the plugs; then you can cut the cable and feed it through the hole.

**4b.** Place 4 rubber feet on the bottom corners of the enclosure.

# 5. MOD THE USB CABLE (OPTIONAL)

**5a.** Cut the USB cord about 3" from the mini-B side (the smaller of the 2 connectors) and set it aside.

**5b.** Feed the long end of the cord through the hole in the enclosure, from the outside.

**5c.** Strip the outer insulation off the cable on each end, and peel away the foil shielding. Strip each of the individual wires inside the cable.

**5d.** On one cable end, place a piece of heat-shrink tubing on each wire. Then connect the matching wires and solder them together. Slip the heat-shrink over the solder joint and use your soldering iron, hot air, or a lighter flame to shrink it around the joint. You can also use electrical tape.

When you've reconnected all the wires, wrap the bundle with electrical tape.

# 6. CONNECT THE CABLE

Plug the the mini-USB plug into the Teensy, and close up the enclosure. Then plug the other end into your computer.

4a

## NOTE
To give the Awesome Button a clean look from the outside, I made the hole for the cable just wide enough for the cable itself, not the plugs. (You can skip this step to save time — just tie a good strain-relief knot in the cable inside the box.)

5a

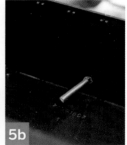

5b

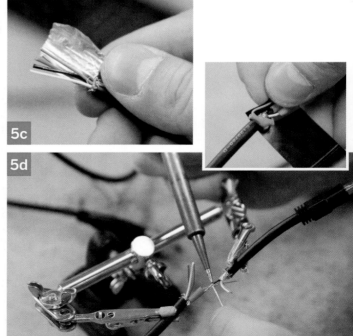

5c

5d

# 7. UPLOAD THE CODE

**7a.** Download and install the Arduino integrated development environment (IDE) from arduino.cc/en/Main/Software.

**7b.** Install Teensy Loader from pjrc.com/teensy/loader.html. Install Teensyduino from pjrc.com/teensy/teensyduino.html. This allows you to use Arduino code and the Arduino IDE to program the Teensy USB microcontroller.

**7c.** Download the Awesome Button code from github.com/mrichardson23/Awesome-Button and open it in the Arduino IDE. If you'd like to make changes to the list of words, add them to the words array in lines 3–5.

**7d.** Make sure that the value of the NUMBER_OF_WORDS constant on line 1 is equal to the number of words in the words array.

**7e.** Under the Tools menu, click Board and choose Teensy 2.0.

**7f.** Under the Tools menu, click USB Type and select Keyboard + Mouse + Joystick.

**7g.** Click the Upload button. A window will appear instructing you to press the button on the Teensy to upload the code. Now, read "Let's Get Unawesome" to learn how to use it. ◾

☑ **TEST BUILDER:**
**Ben Lancaster, MAKE Labs**

Matt Richardson is a contributing editor of MAKE and a Brooklyn-based technophile, maker of things, photographer, and video producer. His work can be found at mattrichardson.com.

6

7f

## Let's Get Unawesome!

Plug the Awesome Button into the USB port of any computer. On a machine that isn't familiar with the device, your operating system may prompt you to hit a particular key so that it can identify the layout of what it thinks is a keyboard. In most cases, you can safely dismiss this dialog box and it won't bother you again.

When you're writing and you're about to type your over-used word, instead slam your hand down on the Awesome Button and it will key in a random synonym so that you don't have to worry about which one to use.

If the word that pops up isn't working for you, you can easily delete it: tap Control-Shift-left arrow on a PC or Option-Shift-left arrow on a Mac to highlight the word and then hit Delete. You could program this key combination into the Awesome Button itself. Or modify the code so that if the button is held down for more than 1 second, it will delete the previous word! Or add a small panel-mount momentary switch to the side of the Awesome Button to do the same thing. Whatever you do, it'll be ... fantastic!

**Make a cheap, high-tech nozzle to eliminate turbulence and create incredible water effects.**

Written and photographed by *Phil Bowie & Larry Cotton*

Gregory Hayes

↗ **TIME:** 4 HOURS (NOZZLE), A WEEKEND (FOUNTAIN)    ↗ **COMPLEXITY:** MODERATE

# Laminar
# *FLOW*
## Water Fountain

Laminar-flow water charms and fascinates. It behaves quite differently from ordinary turbulent water, such as the flow from a faucet or garden hose. A laminar stream is so perfect it could pass for a glass rod. It doesn't splash upon hitting a surface, it will conduct light like a fiber-optic cable, and it's so cohesive, it will enwrap and levitate a smooth sphere, even at a surprising angle to the vertical.

In 2011, we drove 600 miles from our North Carolina homes to Disney's Epcot theme park to study the "Leap Frog" fountain, which chops a laminar stream into arcs, creating impish, cavorting water creatures. We've been obsessed with laminar flow phenomena ever since, joining an online cult of experimenters.

We have achieved laminar flow simply and inexpensively by making a nozzle from a big plastic peanut butter jar, scrub pads, drinking straws, and standard PVC pipe and hose fittings. A fine way to show off its elegant stream is to build a fountain using this nozzle as its heart. It's easy to make, and can produce captivating shapes or even levitate lightweight spheres.

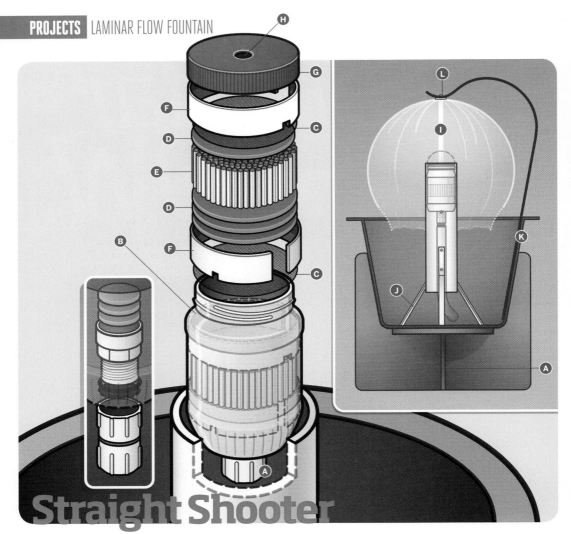

## Straight Shooter

Ordinary water flow is highly turbulent, so it disperses when shot out of a nozzle. Laminar flow is different – aimed by linear stream guides, the water molecules march in one direction like photons from a laser. Creating laminar flow is the key to many technologies, from water jet cutters and Super Soaker squirt guns to boat hulls, sails, and airplane wings.

Laminar-flow fountains are star attractions at Epcot in Orlando, Fla., and Bellagio in Las Vegas. In this home version, you can create amazingly coherent shapes by adding a deflector above the nozzle.

Turbulent water enters the **inlet** Ⓐ via a garden hose.

The cylindrical **nozzle housing** Ⓑ is a plastic peanut butter jar. Its wide diameter reduces water velocity and increases the time the water spends in the nozzle.

**Screens** Ⓒ further reduce water speed, and help distribute flow and reduce flow noise.

**Fiber inserts** Ⓓ made from scrub pads reduce turbulence and minimize eddy formation.

Linear, tubular **stream guides** Ⓔ made of plastic drinking straws divide the flow into multiple straight passages, and their flexibility may further reduce eddies.

**Spacers** Ⓕ retain all these flow components in place.

The nozzle is capped by a **head plate** Ⓖ, whose small **outlet aperture** Ⓗ rapidly contracts the diameter of the flow, speeding it up again. The outlet hole's sharp edges permit water release without introducing new turbulence.

**Laminar-flow water** Ⓘ exits the nozzle in a coherent beam.

In the fountain, the nozzle is mounted vertically on a **nozzle holder** Ⓙ placed in a **basin** Ⓚ.

A small **deflector plate** Ⓛ placed directly in the nozzle's flow diverts it into coherent shapes.

Damien Scogin

# Materials

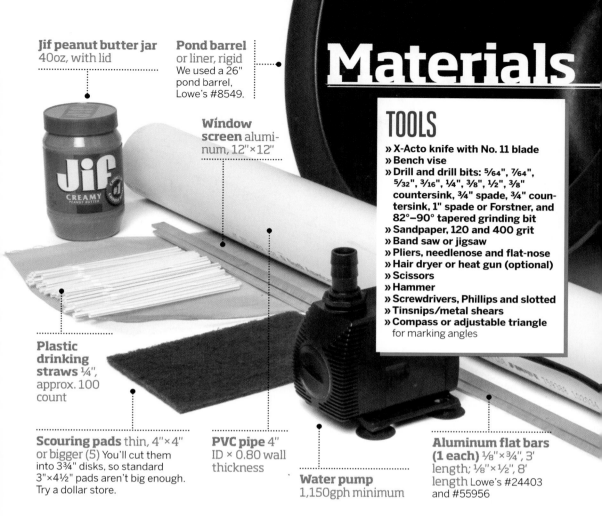

**Jif peanut butter jar** 40oz, with lid

**Pond barrel** or liner, rigid We used a 26" pond barrel, Lowe's #8549.

**Window screen** aluminum, 12"×12"

**Plastic drinking straws** ¼", approx. 100 count

**Scouring pads** thin, 4"×4" or bigger (5) You'll cut them into 3¾" disks, so standard 3"×4½" pads aren't big enough. Try a dollar store.

**PVC pipe** 4" ID × 0.80 wall thickness

**Water pump** 1,150gph minimum

**Aluminum flat bars (1 each)** ⅛"×¾", 3' length; ⅛"×½", 8' length Lowe's #24403 and #55956

## TOOLS

- » X-Acto knife with No. 11 blade
- » Bench vise
- » Drill and drill bits: ⁵⁄₆₄", ⁷⁄₆₄", ⁵⁄₃₂", ³⁄₁₆", ¼", ⅜", ½", ⅜" countersink, ¾" spade, ¾" countersink, 1" spade or Forstner, and 82°–90° tapered grinding bit
- » Sandpaper, 120 and 400 grit
- » Band saw or jigsaw
- » Pliers, needlenose and flat-nose
- » Hair dryer or heat gun (optional)
- » Scissors
- » Hammer
- » Screwdrivers, Phillips and slotted
- » Tinsnips/metal shears
- » Compass or adjustable triangle for marking angles

---

**For the laminar nozzle:**
- » PVC pipe, bell-end sewer & drain type, 4" ID × .080" wall thickness, 2½" length Lowe's item #24140, lowes.com; availability varies, so substitute equivalent parts if necessary.
- » PVC pipe adapter, Schedule 40, ¾" MPT to socket Lasco brand, Lowe's #23856
- » PVC pipe adapter, Schedule 40, hose thread to ¾" FPT Orbit brand, Lowe's #129318
- » O-rings, #17, ⅞" ID × 1¹⁄₁₆" OD × ³⁄₃₂" thick (2) Lowe's #198974
- » In-line valve with hose threads
- » Garden hose with male and female ends

**For a metal nozzle aperture (optional):**
- » Aluminum soda can, empty
- » Acrylic sheet, ¼" thick, about 6"×6"

**For the fountain:**
- » Wood, ¾" thick, scrap
- » Wood dowel, ¾" diameter, 4" length
- » Cyanoacrylate (CA) glue, gel type aka super glue

**For the fountain:**
- » PVC pipe, bell-end sewer & drain type, 4" ID × .080" wall, 18" length Lowe's #24140
- » Pine shelving, 1×12 (nominal), 6' length actually measures ¾"×11¼". Shelving has fewer knots than other 1×12 stock.
- » Wood dowel, ⅞" diameter, 6" length Lowe's #19385 (poplar) or #19424 (oak)
- » Plywood, exterior (treated), ¾"×15"×15"
- » Acrylic sheet, ¼" thick, about 3"×3" You'll cut a 1¼" disk.
- » Furniture glides, hard plastic, non-swivel (4) Lowe's #67022 or similar

- » Weatherstripping, ½"×1¼" maximum section, 1' length
- » Machine screws: 6-32×1" with nuts (3), 6-32×¼" stainless (1), 8-32×½" stainless pan head with nuts (7)
- » Sheet metal screw, #8×1" stainless pan head (1)
- » Wood screws, #10×2" (12)
- » Spray primer and paint (1 can each)
- » Tubing/hose and fittings, various for connecting nozzle, pump, and optional filter (see page 132)
- » Rubber grommet, ¼" ID
- » Grounding electrical plug, 3-prong, 15A, 125V Lowe's #45463 or equivalent
- » Clear plastic fillable ornament ball, 4" diameter (optional) such as Amazon #B000LM65Q0

Gunther Kirsch

# Build your fountain.

## 1. BUILD THE NOZZLE

**1a.** Drill a 1" hole in the center of the bottom of an empty 40oz plastic peanut butter jar.

**1b.** Cut the 4" pipe to make 2 spacers as shown, using a band saw or jigsaw. Bend the tabs inward with needlenose pliers, then press them down with the pliers tips to about 90°.

**1c.** Use scissors to cut the drinking straws into approximately 200 segments about 1¾" long. Cut the scrub pads into 5 disks 3¾" in diameter and 3 disks 1" in diameter. Use shears or tinsnips to cut 2 disks from the aluminum screen, also 3¾" in diameter.

**1d.** Attach the ¾" plumbing fittings to the jar bottom as shown, taking note that the outside fitting has 2 different threads.

**1e.** Fill the jar as shown with the scrub pad disks, screen disks, drinking straw segments, and the 2 spacers you made.

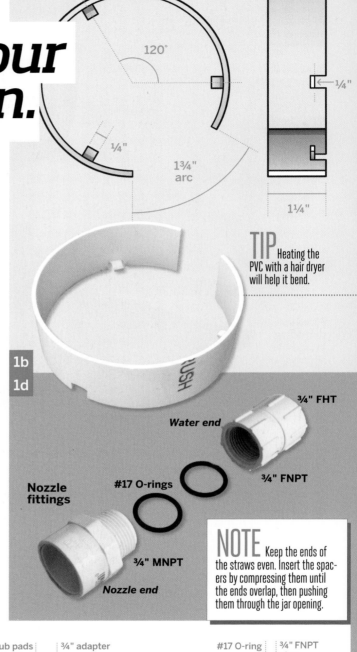

120°

¼"

¼"

1¾" arc

¼"

1¼"

**TIP** Heating the PVC with a hair dryer will help it bend.

1b
1d

¾" FHT

**Water end**

**Nozzle fittings**

#17 O-rings

¾" FNPT

¾" MNPT

**Nozzle end**

**NOTE** Keep the ends of the straws even. Insert the spacers by compressing them until the ends overlap, then pushing them through the jar opening.

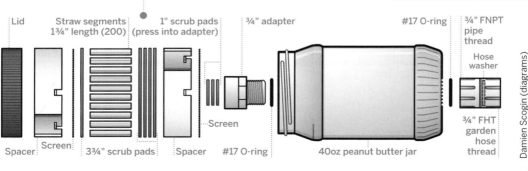

| Lid | Straw segments 1¾" length (200) | 1" scrub pads (press into adapter) | ¾" adapter | | #17 O-ring | ¾" FNPT pipe thread |
|---|---|---|---|---|---|---|
| | | | Screen | | | Hose washer |
| Spacer | Screen | 3¾" scrub pads | Spacer | #17 O-ring | 40oz peanut butter jar | ¾" FHT garden hose thread |

Damien Scogin (diagrams)

# 2. MAKE THE NOZZLE APERTURE

The jar lid must have a perfectly round, sharp-edged ½" hole. We found 2 ways to do this. Method A (potentially cheaper) requires drilling an oversized hole in the lid, then gluing a piece of an aluminum soda can to the underside.

Method B uses only the lid itself — but if the hole is damaged, prepare to eat a lot more peanut butter.

## METHOD A

**2a.** Drill a ¾" hole in the center of the jar lid.

2b

**2b.** With an X-Acto knife, cut a 1"–2" square piece of aluminum from a soda can (0.003" thick). Bend it backward over a ¾" dowel to flatten it, then tape it to a scrap of wood. Using a sharp bit, slowly drill a ⅜" hole in its center. It's OK if it's somewhat crude, because you'll enlarge it to ½".

2c

**2c.** Drill a ½" hole through a piece of ¼" acrylic, backed up with a scrap of wood. Tape the aluminum to the acrylic, keeping the 2 holes aligned.

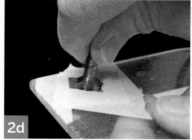
2d

2d

**2d.** Using a sharp ¾" countersinking bit by hand, slowly enlarge the ⅜" hole to match the ½" hole.

Check your progress frequently, and stop when you notice a circular crack in the aluminum.

2e

**2e.** Separate the aluminum from the acrylic and break out the conical aluminum scrap from the hole. You should have a precise ½" hole.

**2f.** With 400-grit paper, gently burnish the hole's inside rim.

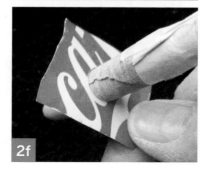
2f

**2g.** Cut off excess aluminum around the hole, leaving about ¼" of material all around. Lightly sand and super-glue the aluminum piece to the lid's underside, keeping the hole centered.

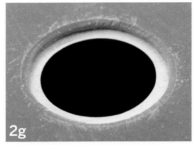

**METHOD B**

**2h.** Drill, from the underside, a ½" hole in the PB jar lid, backing the lid up with a piece of wood. Then grind (or countersink) the hole from the outside to create the important sharp edge.

Keep tools free of material build-up. Use an X-Acto blade and fine sandpaper to eliminate burrs.

Compare the holes made using both methods.

Drilling ½" hole from bottom

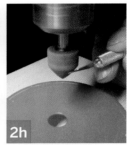

# 3. TEST THE NOZZLE

Screw the lid onto the stuffed nozzle, then test it with a garden hose. We mounted the nozzle to a camera tripod and used an inline valve to adjust the flow.

When the stream smooths out, it should shoot straight up about 12", and when tilted, should be laminar. If your stream isn't laminar, you probably don't have a clean, sharp-edged hole in the PB jar lid.

# 4. MAKE THE FOUNTAIN PARTS

Follow the templates at make projects.com/v/32 to make the fountain parts.

**4a.** Cut and drill the nozzle legs and support arm from the aluminum flat bar stock, then bend them using a bench vise. Cut the deflector from ¼" acrylic, and the nozzle holder from the 4" PVC drain pipe.

**TIP** It's best to cut the pipe by hand. If you must use a bench saw, grip the pipe firmly and approach the blade slowly. Drill the ends of the hose slot with a 1" spade bit, then cut the sides with a jigsaw.

Gregory Hayes (4b)

**4b.** Cut the base legs from 1×12 lumber and the base disk from ¾" plywood, using either a band saw or jigsaw. A drill press helps with the ¼" holes and ⅜" countersinks in the dowel.

If you're not using the 26" pond barrel, then re-size the base parts to fit your fountain basin.

**4c.** Sand, prime, and paint the nozzle holder and the base parts.

# 5. ASSEMBLE THE FOUNTAIN

**5a.** Hammer 3 furniture glides into the bottom of the wood legs, 4" from the outside edges. It helps to drill small pilot holes first.

**5b.** Use 2 of the #10×2" screws to attach the dowel, through any pair of holes, to a leg, flush with the leg's bottom.

**5c.** Arrange the other 2 legs around the dowel, supported by a scrap of wood labeled with 120° angles. Wrap string tightly around this assembly to hold it in position. Don't attach the other 2 legs yet.

**5d.** Position the disk on the legs. Move the legs to match the holes, so that when screws are driven in, they won't split the legs. Keeping the legs vertical, attach the disk with 6 screws. Attach the other 2 legs to the dowels with the remaining 4 screws.

**5e.** Attach the aluminum legs to the nozzle holder, ensuring that the holder sits vertically. Add the nozzle support screws and nuts.

Push the end of a garden hose up through the nozzle holder's slot until it can be attached to the nozzle.

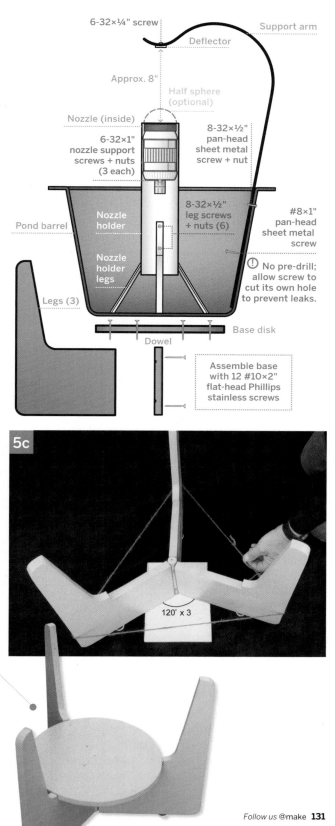

6-32×¼" screw
Support arm
Deflector
Approx. 8"
Half sphere (optional)
Nozzle (inside)
8-32×½" pan-head sheet metal screw + nut
6-32×1" nozzle support screws + nuts (3 each)
Pond barrel
Nozzle holder
8-32×½" leg screws + nuts (6)
#8×1" pan-head sheet metal screw
Nozzle holder legs
⚠ No pre-drill; allow screw to cut its own hole to prevent leaks.
Legs (3)
Base disk
Dowel
Assemble base with 12 #10×2" flat-head Phillips stainless screws

5c
120° × 3

Damien Scogin (diagram)

**5f.** Wrap weatherstrip around, and flush with the top of the nozzle, then push the nozzle into the holder until it stops. Place the assembly in the center of the barrel.

**5g.** Attach the deflector to the support arm with a 6-32×¼" machine screw. If you like, adorn the nozzle top with a clear plastic hemisphere, drilled with a center hole about 1".

**5h.** Test your fountain with the hose; water should deflect into a large, clear, containable dome, or levitate a 3" styrofoam ball. ◪

☑ **TEST BUILDERS:**
**Isabella Ghirann**
**and Brian Melani,**
**MAKE Labs**

5g

5h

Larry Cotton is a semi-retired power-tool designer and part-time community college math instructor. He loves music and musical instruments, computers, birds, electronics, furniture design, and his wife — not necessarily in that order.

Phil Bowie is a lifelong freelance magazine writer with three suspense novels in print. He's on the web at philbowie.com.

# 6. Connect the pump and optional pre-filter.

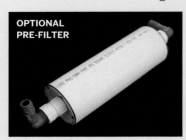
OPTIONAL PRE-FILTER

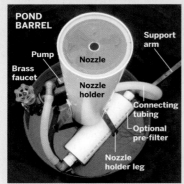
POND BARREL

Pump
Nozzle
Support arm
Brass faucet
Nozzle holder
Connecting tubing
Optional pre-filter
Nozzle holder leg

Many different pumps can be used with this fountain: magnetic drive, direct drive, submersible, in-line, bilge, sump, pond, waterfall, utility, and others. If you have a good water pump on hand, try it with the PB jar nozzle.

If you use a submersible, remove its plug and route the cord through the pond barrel's wall using a sealing grommet. Reattach the plug or buy a new grounded plug, such as Lowe's #45463.

The pump we had on hand (Smart Garden Infinity, 1,150gph), when used with the PB jar nozzle alone, on municipal or well water, produced a nice laminar flow and fit in the pond barrel.

However, we wanted a large, clear dome whose spray could be precisely controlled and recirculated. To do that, we added a 1" brass faucet and a pre-filter between the pump and the nozzle. We wrapped a coarse scrub pad over the faucet inlet to block debris.

You may find that your pump works fine with the nozzle alone. Otherwise, follow the assembly diagram and materials list at makeprojects.com/v/32 to build the pre-filter.

Use kinkless hose or ½" surgical tubing to connect the fittings. For tight tubing bends, you can enclose the bend section in a length of ribbed plastic bilge pump hose to minimize kinking. Hose clamps are usually unnecessary with surgical tubing.

will produce a beautiful, large, clear dome. Position your fountain in a sheltered area away from wind, because domes can assume uncontainable (though interesting) shapes.

Adjust the faucet to control the size of the dome so the water is collected in the basin.

You may want to play with different deflectors, or try levitating different spheres. Add LEDs or other lights for nighttime viewing. Cruise a few of the websites devoted to laminar-flow water features for endless ideas. A particularly good one is laminar.forumotion.com.

And prepare to become addicted.

CAUTIONS Always use a GFCI outlet to connect your pump's power cord, to protect yourself against possible short-circuits caused by water.

Because of possible loss of water from breezes and subsequent pump damage (most pumps can't operate dry), this fountain should not be used unattended.

Gregory Hayes

# World Control PANEL

Written and photographed by **Steve Lodefink**

### Global domination for the young evil genius.

⚡ **TIME:** A FEW WEEKENDS ⚡ **COMPLEXITY:** MODERATE

My son Harlan and his pals love to play "Agents." When he asked me if I could build him a control panel that had "a bunch of switches and random blinking lights," I couldn't have been happier.

I have to admit, I have a bit of a control panel fetish, and my favorite thing about electronics tinkering is making lights blink – which is just about all this device really does.

Gregory Hayes

I immediately imagined some kind of backlit, brushed-metal map of the world, with a radar scanner and a variety of indicator lights and toggle switches that would allow him to communicate with field agents, remotely dispatch weaponized sharks, or detect enemy activity.

The console employs a combination of readymade and custom circuits to achieve various lighting and sound effects. Ours is used for world domination, but the same basic panel would work equally well for tracking the migration of a swarm of monarch butterflies, or detecting unicorns, if that's what you need. Here's how I built it.

## 1. Design the panel.

I had a solid idea of how the perforated map display might work, but I wasn't sure how to achieve my vague "radar scanner" concept. Then I remembered that Evil Mad Science (where else would you buy parts for a world domination panel?) sells a kit to build a

Larson Scanner — you know, that light-chaser effect made famous in the *Knight Rider* and *Battlestar Galactica* TV shows. It's perfect for "scanning the world" before flipping the various function switches. I also ordered amber, blue, and red 3mm diffused LEDs from EMS.

RadioShack's digital sound-recording module sells for $10 and uses nonvolatile memory, so your last recorded sound isn't lost when the power goes away. I added one of these to the panel to serve as a "field communicator."

In keeping with the 1960s sci-fi/spy motif I had in mind, I decided to control everything with a bank of classic metal toggle switches and indicator lights.

## 2. Build the cabinet.

You can use a pre-built enclosure, or build your own. Here's how I made mine, angled toward the operator like an old studio mixer.

For the left and right sides, cut an 11" length of 1"×4" hardwood lumber in half diagonally,

# Materials

» **Switches, toggle, panel mount (7)**
» **Switches, momentary, panel mount: medium (2), any size (1)** for sound buttons and the Larson Scanner mode button
» **EMS Larson Scanner Kit** from evilmadscience.com. Get the one with the 10mm diffused LEDs.
» **Digital recording module, 9V** RadioShack #276-1323, radioshack.com
» **LEDs, 3mm, diffused: blue (25), red (10), amber (10)**
» **LEDs, 5mm: green (4), red (2)** for switch indicators
» **LED, 5mm, flashing, red (2)** the kind with the integrated flasher
» **Resistors, various values (62)** All LEDs in this project need current-limiting resistors. Typical values for different colored LEDs running at 4.5V are: red 180Ω, blue 68Ω, white 68Ω, green 120Ω, and amber 150Ω. But be sure to use resistors that are appropriate for the actual LEDs that you use.
» **Cable ties, small (7)**
» **Heat-shrink tubing (optional)**

*For the flashing backlight:*
» **IC, 555 timer**
» **Transistor, PNP, 2907 type**
» **Resistors: 4.7kΩ (1), 150kΩ (1)**
» **Capacitor, electrolytic, 1µF**
» **Capacitor, ceramic, 0.1mF**
» **LEDs, high brightness: white (7), wide angle: red (3)**
» **Perf board, about 2"×3"**

*For the enclosure:*
» **Hardwood lumber, 1"×4", 26" total length**
» **Aluminum sheet, ⅛"×14½"×10⅛"**
» **Plywood, ⅛"×11"×15½"**
» **Diffusion paper or gels** such as amazon.com #B000265DJU
» **Testors decal paper, clear**
» **Acrylic sealant, clear**
» **Wood screws, small (8)**
» **Wood glue**
» **Battery holder, 3×AA**
» **Hookup wire**

# TOOLS

» **Table saw or miter saw**
» **Router with rabbeting bit (optional)**
» **Spray-mount adhesive**
» **File or belt sander** to deburr and round aluminum edges
» **Center punch** with a sharp tip
» **Hammer**
» **Drill and bits: ¹⁄₁₆", ⅛", ¹³⁄₆₄", ¹¹⁄₃₂", 10mm** You'll also need bits to match your toggles and buttons; ours were about ¼" and ⁹⁄₃₂".
» **Rubber mallet (optional)** to flatten the aluminum if you bend it
» **Sander, random orbit**
» **Sandpaper, 50 and 180 grits**
» **CNC cutter (optional)** I wish I'd used one for all the drilling!
» **Screwdriver, Phillips head**
» **Wire cutters and strippers**
» **Soldering iron and solder**
» **Hot glue gun**
» **Computer and inkjet printer**

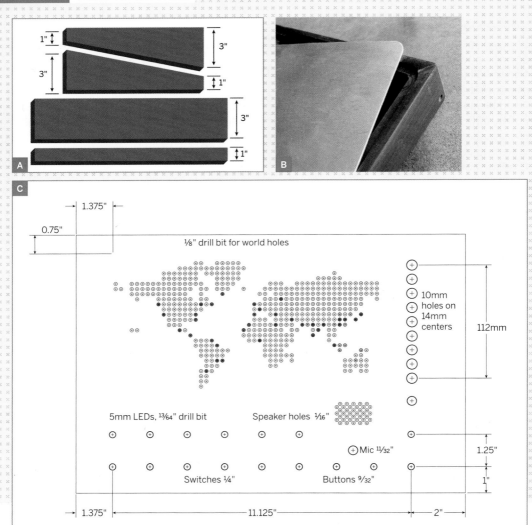

creating 2 pieces, each 1" high at one end and 3" at the other. For the front and rear sides, cut a 15" length straight across to leave a 1" wide piece for the front, and a 3" piece for the rear (**Figure A**).

Miter all 4 sidepieces to 45°, then clamp and glue them like a picture frame. Once the glue has dried, sand the joints flush. (I had to add screws later because glue wouldn't stick to the oily ipê wood that I used — don't use ipê!)

Cut a bottom panel 11"×15½" from ⅛" plywood and attach it with 4 wood screws.

The top panel is a 14½"×10" piece of ⅛" aluminum sheet. I used a router to cut a ⅛"-deep rabbet in the topside of the wooden

frame, to allow the top panel to sit flush. Then I rounded the corners of the aluminum with a 50-grit sanding block to match the radius of the rabbeting bit (**Figure B**).

## 3. Prepare the panel.
Figure out exactly where you want all your controls and displays to go. Once you're satisfied with your layout, mark everything for drilling with a scribe or punch.

I found a "pixel map" of the world online, and scaled it to a size that worked well with my panel dimensions (**Figure C**). I printed it out as a template and glued it to the panel with spray mount. Then I used a center punch and

# UPGRADES

Drilling all the LED holes is an ideal task for a CNC machine! Even when using my table saw as an "anvil," the cumulative force of all this punching caused significant deformation to the panel, and the actual drilling was laborious.

If I were to do it again, I'd try an acid etch process for the panel icons.

hammer to mark each "pixel" for drilling.

Since I would be inserting 3mm LEDs through some of the map holes, I drilled them with a ⅛" bit, which is a little over 3mm.

Then I drilled holes for the Larson Scanner, the sound recorder's speaker grill and mic, the toggle switches and 5mm indicator LEDs, and the buttons that operate the sound recorder and Larson Scanner. You can download my drilling template at makeprojects.com/v/32.

Finally, I drilled 4 small holes near the corners, for screws to secure the panel.

Once all the drilling was done, I pounded the panel flat with a rubber mallet, and used a random-orbit sander to deburr all the holes at once and clean up the panel.

To give the aluminum a nice "brushed" look, I hand sanded it with 180 grit in long, even, horizontal strokes.

Since I was planning to use backlighting effects for the map display, I glued a sheet of photographic diffusion paper to the backside of the map, to act as a rear projection screen. You can just poke holes in the diffusion paper wherever you want to install an LED.

Finally, I printed some free web UI icons ("Brightmix" icons, from opengameart.org)

on a sheet of clear Testors decal paper and applied the appropriate icon above each toggle switch hole. These decals need to be sealed with clear acrylic before use.

## 4. Install toggles and batteries.

Install the toggle switches and indicator lights into their mounting holes. I cinched small zip ties around the base of each indicator LED as spacers so they sit near-flush with the panel. Wire it up so that the main power switch feeds power to the rest of the switches, and each switch powers its own indicator LED.

Power is supplied by a 3×AA battery box mounted to the back panel. Use alligator clips to temporarily connect it to the main power switch. You'll connect and disconnect it many times during assembly and troubleshooting.

## 5. Build the "red alert" flasher circuit.

I wanted a flashing backlight effect to indicate a "global red alert" situation. To achieve the flash, I turned to the good old 555 timer IC. I found a simple 555 flasher circuit on the web (**Figure D**) and built it onto a small piece of perf board (**Figure E**). I used the 555's output to flash 7 white LEDs through a single PNP transistor. I also added 3 bright red 10mm LEDs (these don't flash) to the board. Follow the schematic to build the board.

Leave the LED leads long so you can bend them to direct the light to different areas of the map. Mount the flasher board to the bottom

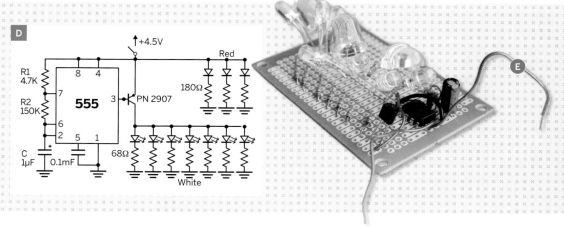

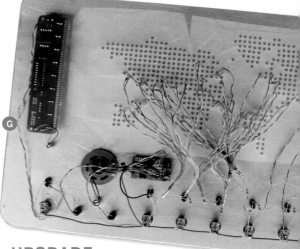

panel with hot glue so it can project its light onto the back of the map. The switch indicator lamp for the "red alert" function is a 5mm red flashing LED.

## 6. Add the map LEDs.

Four of the toggle switches control clusters of different colored LEDs — in my case, blue for major cities, red for Atlantic coastal areas, amber for "Eastern locations," and a single blue LED for Seattle, our home base (Figure C). I also wired a red flashing LED behind the panel, parallel with the Seattle LED, to illuminate the surrounding holes with a flashing "home base" effect. These functions feature green power indicator LEDs above their switches. All of the LEDs in this project require appropriate current-limiting resistors.

To wire the through-hole map lights, I first harvested some twisted-pair wiring from an old network cable I found in the trash. The wire inside these cables is conveniently color-coded and pre-bundled into pairs — great for hand-wiring small electronics projects.

Solder a current-limiting resistor and a length of colored wire pair to each 3mm LED, and then insert the LEDs into the map holes from behind, securing each with a dab of glue (**Figure F**).

Then gather up the similarly colored LED wires and solder the anode wires to their respective power switches, and the cathode wires to ground (**Figure G**).

## 7. Install scanner and sound.

The Larson Scanner is configured to run on 3V

# UPGRADE

If you're so inclined, you could make a printed circuit board to run all the map LEDs, and eliminate much of the hand wiring.

power. Since the World Control Panel operates on 4.5V, you'll have to replace the kit's current-limiting resistors with new 120Ω resistors.

The scanner PCB also has a momentary switch that controls its speed and brightness; I replaced it with a panel-mount button.

Install the sound module by hot-gluing both the speaker and the PCB to the backside of the top panel. I replaced the onboard buttons for recording and playback with a pair of red and black panel-mount momentary buttons. Although the datasheet for the module specifies 9V power, mine worked fine with 4.5V.

Finally, solder the wires to the battery box, and secure the panel with 4 wood screws.

## Take It Further

The World Control Panel is really just a fun box of switches and lights, but the platform begs for further development. Imagine adding USB, wireless data, microcontrollers, BlinkM smart LEDs, and other technology to improve the fun — or to really start monitoring and controlling your global concerns. Let us know what you do with yours at makeprojects.com. ◼

◼ The WCP in action: makeprojects.com/v/32

Steve Lodefink is a software user interface designer, broad-spectrum tinkerer, and promiscuous hobbyist who lives in in Seattle.

# Composting Toilets
## MADE EASY

### Poop in a bucket, save the Earth.

Written and photographed by *Tim Anderson*

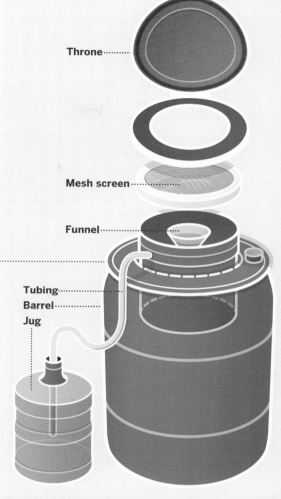

Throne

Mesh screen

Funnel

Tubing
Barrel
Jug

⚡ **TIME:** 1 HOUR   ⚡ **COMPLEXITY:** EASY

Your excrement contains the nutrients to fertilize and grow the food that feeds you. This was known and practiced from the dawn of agriculture until modern times.

Then the use of chemical fertilizers made it possible to discard excrement into sewers instead of returning it to fields. The germ theory of disease made it seem desirable to do so.

Unfortunately, this has resulted in vast amounts of purified drinking water being used to flush soil fertility out into the oceans. In the oceans these excess nutrients cause "dead zones" from algae overgrowth and decomposition, which depletes the oxygen in the water. In these "eutrophic" areas, fish literally drown.

We now know it's possible to eliminate pathogens by composting instead of flushing. We can return to the sustainable practices of ancient times without fear of disease, by using composting toilets. I decided to try it.

### Ancient Roots

In 1905, American agricultural scientist F.H. King traveled to China, Japan, and Korea to study traditional farming practices there. He was very impressed by the absence of flies in those countries. Human excrement ("night soil") and every other material that could decay was immediately collected and sold to farmers. The farmers carried it from the city to the countryside to compost it and use it for fertilizer. Flies had nothing to feed on.

There were stormwater drains, but no sanitary sewers — excrement was far too valuable to flush away into a river. King's classic book *Farmers of Forty Centuries, Or Permanent Agriculture in China, Korea, and Japan* inspired the permaculture movement. It's free online from archive.org, gutenberg.org, and others.

### A Simple Composting System

**Figure A** (page 140) shows a composting barrel, a Luggable Loo bucket toilet (about $20 at REI and other retailers), and a jug-and-funnel urinal.

Damien Scogin

A

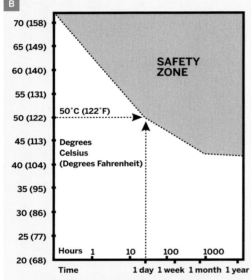

B

One day at 50°C (122°F) is a time and temperature combination yielding total pathogen death for common disease organisms that can be transmitted by humanure. Lower temperatures require longer retention times.

Pathogen death boundaries shown include those for intestinal (enteric) viruses, *Shigella*, *Taenia* (tapeworm), *Vibrio cholera*, *Ascaris* (roundworm), *Salmonella* and *Entamoeba histolytica*. [Source: Feachem, et al., 1980]

I lined the 5-gallon Loo bucket with double paper bags and an inch or two of sawdust on the bottom. Toilet paper and more sawdust go into the Loo with each use. When it's full, you empty it into the composting barrel.

To make the composting barrel, I got a bucket with an easily removed snap-on lid, and cut the bottom off it. Then I cut a hole in the top of a barrel and jammed the cut-off bucket into the hole. I cut a hole in the bucket

lid for airflow, and stapled fine mesh screen over the lid to keep out flies.

After a while I got lazy and decided to skip the Loo. I Craigslisted a hospice throne and perched it over the barrel. I hold a pee bottle in front as an impromptu urine diverting system.

Peeing separately reduces the barrel capacity needed by more than half. It will take one person about a year to fill the barrel.

## Urine Command

Urine is the safest of bodily fluids — typically it's sterile. In most parts of the world it's probably safer to have contact with urine than with the local water supply. Leptospirosis and schistosomiasis can be carried by urine, but if those diseases are in your area, it's still usually better to apply urine to the soil or a compost pile than to flush it into a body of water.

The urine-diverting throne has a funnel in the front of the "drop zone" (as seen on page 139) which carries the urine away to a jug for immediate use as an excellent fertilizer. This greatly reduces the volume of material that goes into the composter. The weatherstripping on the underside of the seat and lid is there to block insects. For a one-way valve, drop a ping-pong ball into the funnel; pee goes in, smells don't come out. A water trap pipe from a sink with mineral oil in the upper part can do the same.

Adding carbohydrate-rich food waste (like bread or rice) to the jug will help the urea in the urine ferment into nitrates rather than volatile ammonia. If you smell ammonia, add more cellulose or carbohydrates. Peeing directly on a bale of straw is a popular solution. Carol Steinfeld's *Liquid Gold* (liquidgold book.com) is an entertaining and informative book about urine as fertilizer.

## What About the Smell?

My humanure barrel smells like damp sawdust. I love showing it to people because they always say, "That doesn't smell bad at all!" Then I get to say, "That's right. My sh*t doesn't stink!"

Add sawdust to your bucket toilet until it smells nice. That's a couple of handfuls of

*The Humanure Handbook*, Joseph Jenkins (chart)

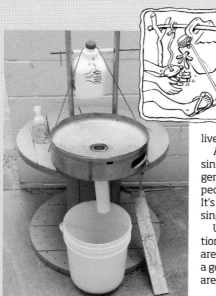

# Hand Washing

The Tippy Tap, a simple foot-operated hand-washing station, can be built anywhere. Put them all over your yard, house, and neighborhood! Hand washing isn't just for compulsive germophobes. Tippytap.org provides statistics on the millions of lives that could be saved if more people washed their hands.

A more elaborate station (left) uses a cut-off keg top as a sink. The foaming hand soap uses a potassium-based detergent that's better for the garden. The cord goes from the foot pedal through a hole in the table, around the neck of the jug. It's amazing how much hand washing can be done with a single gallon of water.

Use plain soap and water for washing, not antiseptic concoctions. Our bodies contain vast numbers of helpful microbes that are necessary for good health. Routine use of antiseptics is not a good practice. It leaves your skin like a petri dish — a vacant area ready to be invaded by opportunistic life forms.

sawdust per use. At that point the ratio of carbon and nitrogen is perfect for the growth of thermogenic aerobic bacteria, which generate heat, $CO_2$, and water. The long-chain nitrogen compounds that make feces stink are no longer being produced.

A year or two after the last addition to a humanure barrel, it will have composted down to one-quarter or less of its original volume and will smell like black dirt. It was the smell of a jar of finished humanure that won me over. That wholesome black-dirt smell was more convincing than any theories or books.

## What About Germs?

*The Humanure Handbook* by Joseph Jenkins (humanurehandbook.com) explains a system using a thermogenic compost pile in a straw-lined corral of old pallets. His table in **Figure B** summarizes the conditions for pathogen elimination. Germs die and become food for harmless bacteria in the hot, damp conditions of the pile.

The World Health Organization provides data about pathogen survival in composting conditions in their "Guidelines for the safe use of wastewater, excreta and greywater" (makezine.com/go/whowaste). At lower temperatures it takes longer to eliminate pathogens. Dry, cold conditions are the least effective.

When your barrel is full, set it aside for the sufficient time for pathogen elimination. (Most people won't get the urge to mess with it prematurely anyhow.)

## Freedom from Flies

When I first made my barrel I used window screen on the bucket lid. One hot day it became obvious I had an insect problem. Big, loud, black flies had gotten into the barrel.

I read the tales of woe from compost toilet users proofing their systems against "insect escape." I made a new fine-mesh lid with no-see-um netting sandwiched between 2 layers of window screen. I added a few handfuls of sawdust to the barrel, put on the new lid, and shook up the barrel. A week later: no insects to be found. Triumph without chemicals!

## Remember

It's no coincidence that our bodily excretions are what plants need. We belong on this planet and fit perfectly with the plants that feed us. So remember, don't put poop in a pipe! ◪

USDA manure and nutrient cycle chemistry: makezine.com/go/manurechem (PDF)

Tim Anderson (mit.edu/robot) is the co-founder of Z Corp. See a hundred more of his projects at instructables.com.

tippytap.org (illustration)

⤢ **TIME:** AN HOUR
⤢ **COMPLEXITY:** EASY

## Build the clever instrument that told sailors their latitude for 200 years.

By **William Gurstelle**
Illustrated by **Damien Scogin**

# Levi ben Gershon
## and the
# Jacob's Staff

In the days before GPS satellites and computers, how did sailors like Christopher Columbus navigate from one port to another? Well, there were several methods.

Dead reckoning was one. The captain would trail a floating line from his ship and record the time it took to play out completely. He then knew roughly how fast the ship was traveling. By combining speed with a compass heading, he could estimate his change in position.

Dead reckoning wasn't very good for long-distance travel. Instead, medieval- and Renaissance-era navigators typically "ran down a latitude." That meant the captain would take the ship to whatever latitude the desired port was on and then steer due east or west until he more or less ran into it.

This simple method required an instrument that would allow the navigator to accurately determine a ship's current latitude. The first navigational instrument that could do so was invented by a now little-known, but immensely gifted 14th-century French mathematician and rabbi named Levi ben Gershon, also known as

Gersonides. He was responsible for a number of important advances in the fields of geometry, trigonometry, logic, and mathematical education. But Rabbi Levi was more than a theoretician. He applied his mathematical discoveries to real-world problems. The most notable example is the Jacob's staff.

Also called a cross-staff, Levi's invention was an instrument with a pair of sliding sights mounted on a horizontal bar bearing a carefully incised geometric scale. By sighting the measured object and the horizon simultaneously and observing the arc subtended between the two, the surveyor now had an accurate method of determining angles and therefore latitude. For more than 200 years, the Jacob's staff was used by European captains to find their way at sea. There was nothing better until the backstaff was invented in 1594 by English sea captain John Davis.

### Make a Jacob's Staff

**1.** Cut the dowels and cut a 45° angle onto one end of each 4½" dowel.

To make the sighting crosspiece or "transom," butt the short dowels against either side

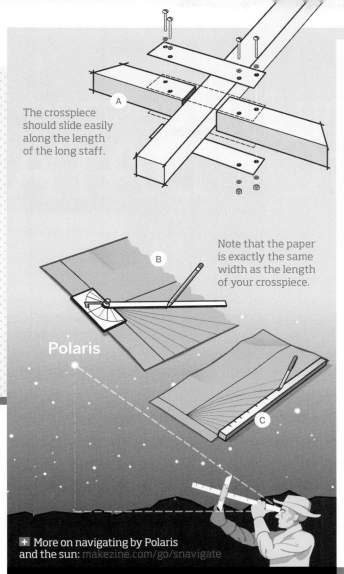

The crosspiece should slide easily along the length of the long staff.

A

Note that the paper is exactly the same width as the length of your crosspiece.

B

Polaris

C

+ More on navigating by Polaris and the sun: makezine.com/go/snavigate

## Go Navigate

After dark, go outside and hold the Jacob's staff as shown. Slide the sighting crosspiece until the top edge aligns with the North Star and the bottom edge aligns with the horizon. Read the elevation angle of the North Star on the scale marked on the staff; this angle is your current latitude!

If you travel south, the North Star will appear to sink in the sky, so you'll have to extend the crosspiece to find your current latitude. If you travel north, then vice versa.

Experienced navigators can also use the Jacob's staff to take noon sun sights (don't stare at the sun!) and to determine the height or distance of distant objects. Bon voyage!

of the long dowel as shown, in line with each other, with their longest sides facing the same way. Center the brass strips over them, drill four ³⁄₁₆" holes completely through the brass and short dowels, and fasten them with the machine screws, washers, and nuts (**Figure A**).

**2.** Mark the paper with a vertical line 2" from one of the shorter edges, and a horizontal centerline down the long dimension of the sheet.

Align your protractor's bottom edge with the vertical line and place the protractor's center point at the intersection of the 2 lines.

With a pencil, draw lines at 70°, 60°, 50°, 40°, 30°, and 20°, radiating outward from the protractor's center point to the paper's bottom edge (**Figure B**).

**3.** Place the staff next to the paper's bottom edge. Align one end of the staff with the vertical line. Mark the staff with the corresponding degrees where the lines intersect the paper's edge (**Figure C**). Slide the crosspiece onto the staff so its long side faces the high end of the scale. This is the eye-end.

Congratulations, your Jacob's staff is complete and you're ready to navigate! ◪

William Gurstelle is a contributing editor of MAKE. The new and improved edition of his book *Backyard Ballistics* has just been released.

# Making Synthesized MUSIC *from your* DATA

By *Forrest M. Mims III*

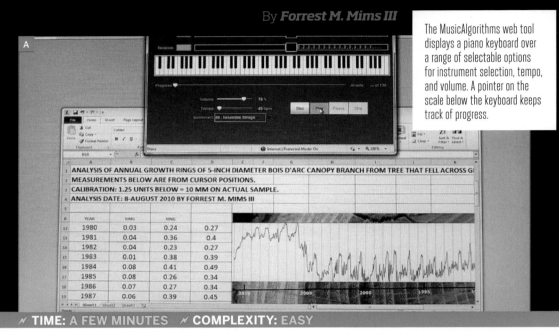

The MusicAlgorithms web tool displays a piano keyboard over a range of selectable options for instrument selection, tempo, and volume. A pointer on the scale below the keyboard keeps track of progress.

| | ANALYSIS OF ANNUAL GROWTH RINGS OF 5-INCH DIAMETER BOIS D'ARC CANOPY BRANCH FROM TREE THAT FELL ACROSS Gi |
| 2 | MEASUREMENTS BELOW ARE FROM CURSOR POSITIONS. |
| 3 | CALIBRATION: 1.25 UNITS BELOW = 10 MM ON ACTUAL SAMPLE. |
| 4 | ANALYSIS DATE: 8-AUGUST 2010 BY FORREST M. MIMS III |

| | YEAR | RING | RING | |
|---|------|------|------|------|
| 12 | 1980 | 0.03 | 0.24 | 0.27 |
| 13 | 1981 | 0.04 | 0.36 | 0.4 |
| 14 | 1982 | 0.04 | 0.23 | 0.27 |
| 15 | 1983 | 0.01 | 0.38 | 0.39 |
| 16 | 1984 | 0.08 | 0.41 | 0.49 |
| 17 | 1985 | 0.08 | 0.26 | 0.34 |
| 18 | 1986 | 0.07 | 0.27 | 0.34 |
| 19 | 1987 | 0.06 | 0.39 | 0.45 |

**TIME:** A FEW MINUTES    **COMPLEXITY:** EASY

Do you enjoy the sounds that wind chimes extract from a soft breeze? How about the gentle splashing of raindrops, the soothing sound of falling water, or the roar of surf?

These appealing natural sounds provide only a hint of the vast range of musical compositions hidden away in many kinds of data. Lately I've been having lots of fun transforming data I've collected into natural music and posting the results on youtube. com/fmims. Various methods for making sounds from data are available. Let's use them to convert data into music.

## Using Mathematica to Create Music

George Hrabovsky is an amateur physicist who uses Wolfram Mathematica software

(wolfram.com/mathematica) in his theoretical research. His praise of Mathematica was so persuasive that I eventually bought the program, and it's where I first went when trying to transform data into music.

Among Mathematica's astonishing range of features is the ability to convert numbers into synthesized musical notes representing a variety of instruments. Mathematica's Music Package can convert data into representative audio frequencies and much more. If you're into programming, it's a highly flexible tool for transforming data into music.

## MusicAlgorithms

Jonathan Middleton is assistant professor of theory and composition in the music department at Eastern Washington University, where he teaches composition, orchestration, and computer music. While exploring ways to transform data into music, I discovered Dr. Middleton's MusicAlgorithms website, which he developed with assistance from Andrew Cobb, Michael Henry, Robert Lyon, and Ian Siemer with sponsorship from the Northwest Academic Computing Consortium.

The homepage states that, "Here, the algorithmic process is used in a creative context so that users can convert sequences of numbers into sounds." That single sentence hooked me into the MusicAlgorithms site for a week while I transformed some of my data into an amazing variety of intriguing musical "compositions."

## How to Use MusicAlgorithms

MusicAlgorithms requires a Java-enabled computer. Transforming a string of numbers into music is simple; you can either type or paste a series of numbers into the program. Here's a quick way to learn to use the site:

**1.** From the homepage (musicalgorithms.ewu. edu), click the Compose button. Then click "Import your own numbers" to enter the data input page.

**2.** In the Algorithm box, enter into window A the numbers 1 through 10 (press the Enter key after each number).

**3.** Ignore checkboxes B, C, and D and click the Get Algorithm Output button.

**4.** In the Pitch box, click the Scale Values button to normalize the numbers 1 to 10 that you entered in window A into the piano scale of 0 to 88 (1 = 0, 2 = 9, 3 = 19, ... 10 = 88). These numbers will appear in the adjacent Derived Pitch Values window.

**5.** Skip the Duration box (for now). In the Compose box, click the Play button.

**6.** A MIDI Player window will open, showing a piano keyboard over buttons for Step and Play and options for Volume, Tempo, and Instrument (**Figure A**).

Click the Play button to hear the 10 notes you have composed. Keep playing these notes while using the sliders to adjust the volume and tempo. Then let the fun begin by selecting from the pull-down menu of 128 synthesized instruments and sounds. Soon you'll be ready to compose music from real data.

## Finding Data for MusicAlgorithms

If you're an amateur weather watcher, you probably have plenty of numbers to transform into music. For example, MusicAlgorithms will convert a year of your daily minimum and maximum temperatures into a remarkable audio experience that will provide an entirely new way to appreciate your data. If you have no scientific data, try converting your daily expenses or bank balance into music. You might be surprised by what you hear.

A goldmine of data is scattered across the web. For example, my local National Weather Service station near San Antonio, Texas, provides monthly and annual precipitation data since 1871 and temperature since 1885. Converting these data into music provides an entirely new way to better appreciate seasonal temperature cycles and even cold fronts, El Niños, and droughts.

The U.S. Geological Survey provides data on stream flow. Many NASA and NOAA sites are filled with data. Other data sources include the U.S. census, stock market statistics, commodity prices, grocery store price lists, traffic counts on major highways, and so forth.

## Sample MusicAlgorithms

I've posted several videos of MusicAlgorithms based on my data. These will give you a good idea of the amazing variety of sounds you can produce from data that ordinarily are depicted only as dots, lines, or bars on charts.

### 1. ONE YEAR OF SOLAR NOON UV-B DATA

Since 1988, I've measured the sun's ultraviolet radiation from a field in Texas at solar noon on days when clouds didn't block the sun. In this YouTube clip (youtu.be/VsRCrh6XWog), the UV-B intensity at noon on each of the 170 days during 2010 in which the UV-B could be measured in **Figure B** (page 146) is transformed into representative musical notes.

Each musical note is accompanied by a 360° fisheye image of the sky made when the UV-B was measured. Low UV-B levels during

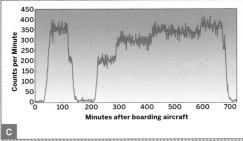

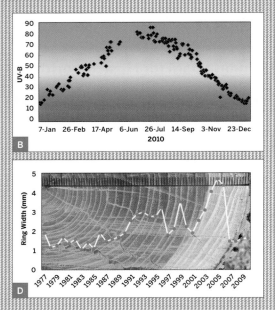

**B.** Weather data, like these solar UV measurements, are easily converted into representative sequences of musical tones.

**C.** The cosmic ray background count measured by a Geiger counter increases with altitude. The altitude changes of an aircraft flying from San Antonio to Zurich can be heard as distinct changes in pitch, proportional to altitude.

**D.** Decades of tree growth can be converted into music based on the precipitation-modulated width of annual growth rings. The bois d'arc tree shown here began growth in 1977.

winter are indicated by low pitch tones, and the high UV-B levels during summer by high pitch tones. Variations in the steady increase and then decrease in UV-B during the year are caused by clouds near the sun, haze, and changes in the ozone layer.

## 2. THE COSMIC RAY BACKGROUND COUNT

In this video (youtu.be/bAKdaYumlq4), the cosmic ray background count on a flight from San Antonio, Texas to Zurich, Switzerland (**Figure C**), is transformed into an audio composition in which the frequency of tones represents altitude. A typical Geiger counter measures around 11 counts per minute (CPM) at the ground and several hundred CPM at altitudes of 35,000 feet or more.

## 3. TREE RINGS TO SYMPHONIC STRINGS

MusicAlgorithms can convert the widths of annual growth rings in trees into a tune in which wide rings from wet years have a higher pitch than thin rings from dry years. This composition uses ring data from a tree at my place downed by a flood in 2010 (**Figure D**).

A tree produces one growth ring each year. The light-colored spring growth is called early wood, the darker summer-fall growth is late wood. In this video (youtu.be/l2g3scrcg20), the width of the early and late wood in each ring was measured and then played in this sequence: *early wood note/late wood note/ sum of early and late wood notes/rest interval to separate the rings.* This clip also includes the data plot I used to form the composition.

## Going Further

Here I've simply transformed strings of numbers into music. MusicAlgorithms lets you do much more. You can select from a variety of mathematical functions and then enter the pitch range and the duration of each tone. These functions include the mathematical constants pi ($\pi$), phi ($\varphi$), and $e$, exponents, the Fibonacci sequence, Pascal's triangle, Markov chains, and even a chaos algorithm and DNA sequences.

If you're interested in synthesized music, you can't go wrong exploring these features. Just block out some time. MusicAlgorithms is the most addictive website I've ever visited. ◪

Forrest M. Mims III (forrestmims.org), an amateur scientist and Rolex Award winner, was named by *Discover* magazine as one of the "50 Best Brains in Science." His books have sold more than 7 million copies.

# 1.2.3. Fast Toy Wood Car

By **Ed Lewis** Illustrations by **Julie West**

**LOTS OF MY FRIENDS HAVE KIDS,** and that means lots of birthdays. I wanted to have a custom present that's easy to make and has lots of room to play, in terms of design.

A toy car fits perfectly. So I can build cars and make kids happy? Win-win!

> **YOU WILL NEED:** Plywood, ¼", 11"×14" or more » Bolts, ⁵⁄₁₆, 4" (2) » **Locknuts, ⁵⁄₁₆" (2)** » Spacers, ½" (4) » **Inline skate wheels with bearings (4)** » Wood glue » **Laser cutter or jigsaw, router, or coping saw, and drill with ⁵⁄₁₆" bit** » Clamps » **Cutting templates**

## 1. Cut the plywood.
Download the templates from makeprojects. com/v/32 and use them to cut the plywood body. Use a laser cutter, or cut with hand or power tools. Sand edges.

## 2. Assemble.
Run each bolt through a wheel, a spacer, the car body, another spacer, another wheel, and a locknut to cap it off. (Remove the wheel's internal spacer if necessary.) » The car is ready to roll! If you want, change a layer or two, or even redesign the whole thing.

## 3. Glue.
Take the car body apart and apply wood glue between the layers. Reassemble, clamp, and let dry. » You now have a toy car that's ready for tons of abuse. It can go very, very fast. Little kids will have no problem moving it around, and bigger kids will enjoy whipping it off ramps to see how it performs.

## Going Further
There's lots of room for customization. Make the body profile realistic or more abstract. Give the car a front and back, or make it symmetrical. Play with the wheel size and the distance between the front and back wheels. Stain or paint can liven your car up, as well as extra details such as names or stickers. Make your car what you want it to be! ◪

Ed Lewis lives in Oakland, Calif., with his wife, two sons, two cats, and a shed full of tools.

# Screwy Light

**Use throwaway 3D movie glasses to experiment with linear and circular polarized light.**

Written and photographed by **Donald E. Simanek**

So you've just seen a 3D movie. I hope you saved those RealD polarizing glasses you paid for. If not, ask at the box office if you could have a few pairs that have been used and will be recycled.

The RealD 3D movie process uses *circular polarization*, unlike the 3D movies of the 50s that were presented using *linear polarization*. If you're into 3D photography and you project your pictures onto a screen, you've probably used linear polarizing glasses. Both types of glasses also have other uses, as we shall see. One advantage of the modern glasses for experimentation is that they can be used as either linear or circular polarizers.

Let's demonstrate some of the surprising effects of polarizing glasses, without digressing deeply into the physical explanations. (For that, see my article "Experiments with Polarized Light" at makezine.com/go/expolar.)

## Circular Polarized Light

**The spooky eye patch.** Put on the RealD glasses. Look at your reflection in a mirror. Now close one eye. In the mirror, one of the polarizing filters appears black — the one over your open eye. You can clearly see your closed eye in the mirror (**Figure A**). Think about it. How can your open eye see through the darkened polarizing filter?

Go ahead, open both eyes, then close the other one — but predict what will happen before you do it.

Predict what you'd see if, while wearing the glasses, you looked at another person also wearing glasses. Then what would you see if you closed your right eye? Would you then see the other person's left or right eye?

**Shiny things.** Place a coin or other shiny object on the table, and look at it through the circularly polarizing glasses. It looks normal, doesn't it? Then remove the glasses and place one of the circular polarizers directly on top of the coin. Now the coin looks dark, black, or maybe purple (**Figure B**). Flip the polarizer over and note any difference in appearance of the coin. Substitute a pocket mirror for the coin. Try some crumpled metal foil.

**Fig. A:** Which picture represents what you will see when you look in a mirror?

**Fig. B:** Why does the coin seen through the circular polarizer appear dark while the quarter, seen directly, doesn't?

**Fig. C:** A circular polarizer is a sandwich of a linear polarizer and a quarter-wave retardation layer. In this diagram, left circularly polarized light passed by the left circular polarizer becomes right circularly polarized light upon specular reflection, and therefore can't pass back through the left circular polarizer. Light scattered from nonreflective surfaces is depolarized, so some of it passes through the polarizer.

# How It Works

» Each lens has a filter consisting of a linear polarizing sheet sandwiched with a quarter-wave plastic retarding sheet. The axes of the polarizers and retarders are aligned so that one eye's filter passes only left circularly polarized light, while the other passes only right circularly polarized light. In both cases, the linear polarizers are nearest the eyes.

» The light from the theater's silvered screen is made up of right-handed circularly polarized light intended for one eye, and left-handed circularly polarized light for the other eye. The filters on the glasses select one and reject the other, so each eye sees only the picture intended for it. The orientation of the axis of the linear polarizing layers is irrelevant for the purpose of modern 3D movies. For more on 3D movie technology, see my article "Making Movies Three-Dimensional" at makezine.com/go/mm3d.

» The bottom line: when you wear these 3D glasses the normal way, you're looking at light that was circularly polarized, then passed through a wave-retarding plate, then through a linear polarizer, so that linearly polarized light reaches your eyes (**Figure C**). For future reference you might want to label your glasses with a "P" inside the frame, for linear polarizer, and an "R" outside, for retarding plate-

## Materials

» **RealD 3D movie glasses (1 or more pairs)**
» **Mirror**
» **Coin**
» **LCD display screen** like a laptop or LCD TV
» **Cellophane** from product wrappers, such as CDs or DVDs
» **Cellophane tape**

A

B

C

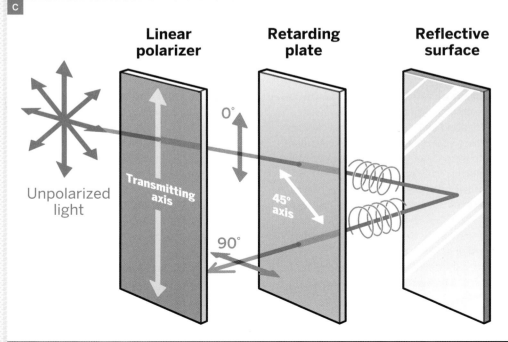

**Linear polarizer**

**Retarding plate**

**Reflective surface**

Unpolarized light

Transmitting axis

0°

90°

45° axis

**Fig. D:** Glasses block light.

**Fig. E:** Glasses transmit light.

**Opposite**

**Fig. F:** Layers of cellophane between parallel polarizers. The cellophane axis is at 45° to the polarizers.

**Fig. G:** The same layers of cellophane between crossed polarizers. One polarizer has been rotated 90° but the cellophane axis is still at 45° to the polarizers.

## Linear Polarized Light

When you look through the 3D glasses backwards, with the retarding plate nearest your eye, the retarder does not affect what you see, and the combination acts as a linear polarizer for incoming light. In this way, you can do all the standard textbook experiments intended for linear polarizers.

**Flip-flop the glasses.** Put the glasses in front of your eyes, but with the earpieces forward (not on your ears). Repeat the previous experiments. Explain them all. It helps to remove the earpieces entirely for some of these experiments, for they just get in the way. (You might also want to cut them apart at the nosepiece with heavy shears for some experiments.)

**Polarization by reflection.** Much of the light we see in everyday life is partially linearly polarized, such as reflections from shiny surfaces. Using your glasses backwards (polarizing side away from your eyes), look at shiny surfaces from different angles, and with different polarizer rotation. You'll notice that at one orientation of the polarizers, reflections from glass at about 56° to the surface (Brewster's angle) are strongly blocked. In this situation, the polarizer's axis is oriented parallel to the reflecting surface. Mark this axis direction on the glasses' frames for future reference.

**Sunglasses.** Glasses for 3D movies are not suitable for use as sunglasses. Polarizing sunglasses have their polarizing axes horizontal, to block reflection from shiny floors, roadways, and water surfaces. Movie glasses have their polarization axes vertical, so even if worn backward they won't work as sunglasses.

**The light from your computer screen is strange.** And it's not just because of the websites you visit. If you have a computer with a flat-screen liquid crystal display (LCD), turn it on. Open your word processor to display the "writer's block" screen (a pure white page).

Now hold the polarizing glasses with the earpieces toward the screen. Rotate the glasses, and you'll find one position where the screen appears black when seen through the glasses (**Figures D and E**). The polarization axis of the glasses will probably then be at 45° to the vertical, but this depends on your brand of computer, and may not be the same for all glasses. Try various glasses with high-definition LCD TVs and other LCD screens.

Now you know why you can't see the digital display in your automobile when you're wearing polarizing sunglasses. Clash of technologies.

## Colors and Cameras

**Creating colors.** While wearing your polarizing glasses, place a piece of cellophane over the computer screen, and rotate it to different positions. In two orientations it will show strong color. Several layers, or different cellophane thicknesses, will show different colors (**Figure F**). Thicker layers are more pastel. Crumpled cellophane may produce an abstract work of color art.

When you rotate the polarizing glasses (take them off first), each color shifts to its complementary color: red to green, yellow to violet, and so on (**Figure G**). You may have better results holding the glasses backwards, so the polarizing layer is away from your eyes. These pictures were taken with polarized light from a computer screen polarized at 45° to the vertical and a linear polarizer in front of the camera lens.

**Be artistic.** Creative types might use this phenomenon for making works of art. In the 19th century, European craftsman made miniature scenes from carefully cut layers of thin crystals sandwiched between polarizing sheets. A sheet of aluminum foil (smooth or crumpled) can be a background for such a scene, using just one polarizer in front of everything. Be inventive.

**Polarizers for cameras.** If you're "cheap," you can use either filter from 3D movie glasses as a polarizer in front of a camera lens by aiming the polarizing side away from the camera (toward the subject). This is useful for eliminating glare from shiny surfaces, increasing the color saturation of shiny leaves, and darkening blue skies for dramatic effect.

In fact, most digital camera manufacturers specify that only circular polarizers be used for this purpose, not linear polarizers. This is because digital camera autofocus and auto-exposure systems usually have an internal mirror to deflect and monitor a fraction of the incoming light, and mirrors polarize the light. So a linear polarizer could, in certain orientations, compromise the accuracy of these important systems.

The incoming light from the scene passes through the linear polarizer, which does its job of reducing reflections. Then the light passes through the retarding plate that converts it to circular polarization, which, being unbiased laterally, cannot confuse the automatic camera systems. Your inexpensive 3D glasses aren't optical quality, so they'll introduce some slight degradation of resolution and sharpness of the image. But you may not even notice the difference. ◪

⊞ To learn more about polarization, see my article "Experiments with Polarized Light" at makezine.com/go/expolar.

For a mathematical treatment of polarization theory, and lots more experiments you can do at home, see makezine.com/go/umichpolar (PDF). (Your RealD glasses provide the equivalent of the raw materials specified in this document.)

Donald Simanek is an emeritus professor of physics at Lock Haven University of Pennsylvania. He writes about science, pseudoscience, and humor at www.lhup.edu/~dsimanek.

# Little **Big** LAMP

**Add bright lighting to your space with powerful LEDs housed in PVC.**

Written and photographed by
*Charles Platt*

✗ **TIME:** 8 HOURS   ✗ **COMPLEXITY:** MODERATE

The most popular item I ever built for MAKE just happened to be the simplest: an LED desk lamp. This was in Volume 08, in 2006, when white LEDs were a hot new product. The most powerful ones I could find were 1cm in diameter, rated to deliver 100,000 millicandles (mcd). The light wasn't exactly white – it had a freaky purplish hue. But I liked the weird color, because it showed we were early adopters of cutting-edge illumination!

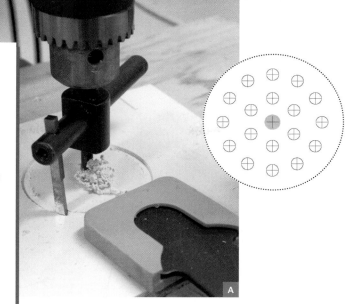

A

## Materials

- » **Power supply, 12V DC** such as 12V/1A AC adapter
- » **LEDs, high brightness, 5mm, 3.3V forward voltage (18)** I used "5-chip" LEDs, 60° dispersion, 100mA. Typically $15–$20 on eBay.
- » **Resistors: 4.7kΩ, ¼W (1); 10kΩ, ¼W (1); 680Ω, ¼W (1)**
- » **Linear potentiometer, 100kΩ**
- » **Capacitors, electrolytic, 10µF (2)**
- » **Capacitors, ceramic, 0.1µF (2)**
- » **Diode, 1N914 signal**
- » **IC, 555 timer, TTL type**
- » **IC, Darlington array, NTE2013 or ULN2003**
- » **Round PCB kit** RadioShack #276-004, radioshack.com
- » **Epoxy, 5-minute, clear**
- » **Knob, ¼" shaft**
- » **Solid-core wire, 22AWG** in red and other color(s)
- » **PVC reducer, white, 2"-to-1½"**
- » **PVC coupling, white, Schedule 40, 2"**
- » **PVC water pipe, white, ½", 2' length**
- » **Extension spring, 3' long, ½" diameter, tempered steel** I used McMaster-Carr #9664K55, mcmaster.com.
- » **ABS plastic sheet, white, ⅛"×12"×12"** or plywood

## TOOLS

- » **Adjustable circle hole cutter** or a small handsaw and sandpaper
- » **Breadboard, 6"**
- » **Awl**
- » **Drill press and bits: ¹⁄₁₆", ⅛", #9, ²¹⁄₆₄", and ¾" or ⅞" Forstner** The Forstner bit should match your ½" pipe's outer diameter, usually 0.84", which is about ⅞" or 22mm.
- » **Heat gun**
- » **Jumper wires**
- » **Multimeter**
- » **Pliers**
- » **Soldering iron and solder**
- » **Wire cutter/stripper**

Recently I wondered if I could downsize and upgrade the original lamp. So I took a fresh look at those traditional, through-hole, single-component LED "indicators" (as they're properly known). The 5mm ones are now a lot more powerful than the 1cm type I used in the past. Some of them, known as "5-chip," have 5 light-emitting elements squeezed into one 5mm package, sucking down 100mA of forward current at around 3.3V DC. They're still rated at approximately 100,000mcd, but 6 years ago, the ones I used were rated for only 20° of beam spread.

Today's 5-chip LEDs claim a spread of 60°. Does that mean they're 3 times as bright? No, they should be 9 times as bright, because the light is delivered over a two-dimensional area! Since I used 72 of the big old ones, and the new ones should be 9 times as bright, I would only need 8 to get the same illumination. But why not go for greater output?

Of course you can still obtain traditional-style 5mm LEDs. RadioShack, for instance, offers them as part #276-017. Since they use exactly the same 3.3V DC as the 5-chip variety, you can substitute them in this project without changing the circuit.

### Fabrication Choices

I opted for a 12V DC power supply, to make the lamp function in motor homes, where LEDs are ideal to conserve power. For use with 115V AC, you need an adapter that delivers 12V DC at 1 amp. Instead of adding a series resistor with each LED, the most efficient way to power them from 12V DC is by series-wiring them in threes. This means you need 10V DC for each set.

How to get 10V from 12V power? Pulse-width modulation is the way to go. You send a stream of pulses, too rapid for the eye to see, and vary the gaps between them to limit the average current. If you add a potentiometer, it can act as a dimmer. Only a few electronic parts are needed. Note that if you use old-style low-power LEDs, your AC adapter can be down-rated to 300mA, which should cost less.

How to build the actual lamp? I decided to use PVC plumbing supplies. For the additional pieces that hold everything together, I chose ⅛" white ABS plastic, but you can use plywood if you prefer.

## 1. Make the lamp head.

Download the template from makeprojects. com/v/32. It shows the layout of the LEDs you'll fit in the large end of the PVC reducer, which forms the head of the lamp.

Use an adjustable hole cutter to cut a circle of ABS or plywood the same size as the template in **Figure A** (page 153). This tool is $10 on eBay, but to use it safely, you need a drill press, as shown here. Otherwise, you can use a handsaw and sand the corners to make a circle.

Tape your template to the circle, and use an awl to poke through the center of each hole (**Figure B**). Remove the paper and pilot-drill through each indentation with a 1⁄16" bit, then drill with a #9 bit. This is the perfect size for the LEDs to push-fit into the holes, so no glue is needed.

## 2. Install the LEDs.

Trim the leads of the LEDs. Make the long leads ½" and the shorter ones ¼", so you can still tell them apart. Push the LEDs into the holes, noting the short and long leads and being very careful to get the polarity right, as shown in **Figure C**.

Solder the LED leads together and add wires as shown in **Figure D**. The red wire powers the positive sides of all the LEDs. The other wires are negative; their insulation colors are arbitrary. I used a separate wire for each group of 3 LEDs in case I might want to light some of them selectively in the future. The front side of the assembly is shown in **Figure E**.

To test the LED assembly, attach the positive side of a 12V DC power supply to the red wire, and the negative side of the supply to your 680Ω resistor. Touch the free end of the resistor very briefly to each negative wire leading to the LEDs. They should light up in threes. If you made a polarity error, the resistor should protect you from burning anything out.

Epoxy the LED plate into the wide end of the PVC reducer to form the head of the lamp. Set it aside to harden.

## 3. Bend the lamp neck.

PVC water pipe has ugly text printed on it, so you'll need to sand it off. Alternatively you can use a solvent, but it may dissolve the PVC or smear the ink.

So-called ½" water pipe varies a lot in internal diameter. First measure your pipe, then order a 3' spring that will fit inside it. The spring will prevent the pipe from kinking when you bend it.

Slip the spring through the pipe (**Figure F**) and wave a heat gun to and fro along the section that you want to bend, while rotating it. Keep the heat gun moving, and be patient. Eventually the pipe will soften, and can be bent into a curve (**Figure G**).

When you've finished, remove the spring and saw the pipe to the size you want for the lamp neck.

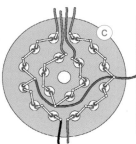

## 4. Make the centering plates.

Cut another 2 circular plates to fit in the narrow end of your lamp head, and use a Forstner bit to cut holes in their centers to fit the exterior diameter of your ½" pipe. Epoxy the plates to the lamp neck, as shown in **Figure H**. Thread the wires through the neck, then epoxy the neck assembly into the head of the lamp.

Make 2 more circular plates to fit in the PVC coupling, for the base of the lamp, and cut holes in their centers to accept the ½" pipe (**Figure I**). Glue the plates into the base: one halfway down, and one at the very top. Then glue the neck into the plates, leaving plenty of room at the bottom for your circuit board and potentiometer.

## 5. Breadboard the circuit.

The schematic diagram (**Figure J**) is config-ured to match your breadboard; you can download a full-sized version from make projects.com/v/32. For testing purposes, insert a 10µF capacitor for C3.

Using a 12V DC power supply, start with only the components around the 555 timer that are shown in the top half of the sche-matic, and attach a single test LED between R3 and negative ground. The potentiometer should now adjust the flashing speed of the LED. If not, you made a wiring error.

Remove your test LED and now use R3 to connect the 555 timer to IC2, the Darlington array, as shown in the schematic. Darlingtons don't source current, they sink current, so connect the red wire from your LED array permanently to your 12V DC source. Then run the negative return wires into the right-hand pins of the Darlington chip. Its left-hand pins are all driven by the 555 timer. The pins labeled "NC" have no connections.

**Figure K** (page 156) shows the bread-boarded circuit. If the potentiometer still makes the lights flash faster and slower, all is good.

Remove the 10µF capacitor you used for C3 and substitute a 0.1µF capacitor so that the lights will flash fast enough to exceed your persistence of vision. Resistor R2 makes sure

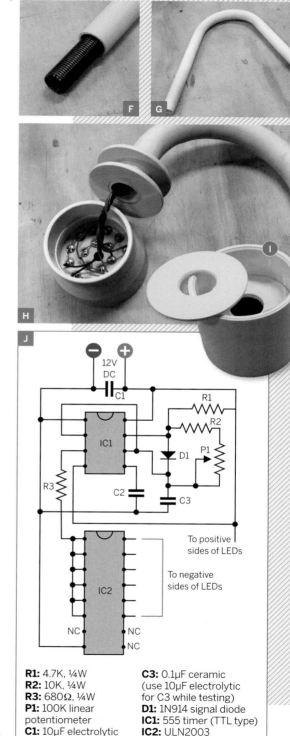

**R1:** 4.7K, ¼W
**R2:** 10K, ¼W
**R3:** 680Ω, ¼W
**P1:** 100K linear potentiometer
**C1:** 10µF electrolytic
**C2:** 0.1µF ceramic

**C3:** 0.1µF ceramic (use 10µF electrolytic for C3 while testing)
**D1:** 1N914 signal diode
**IC1:** 555 timer (TTL type)
**IC2:** ULN2003 Darlington array

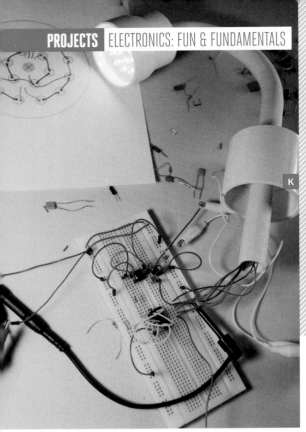

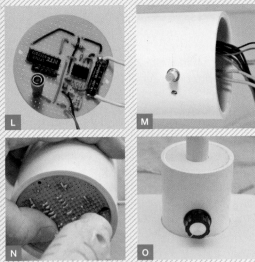

there's always some "off" time in the pulse train, even when the potentiometer is at the end of its range. This way, the LEDs won't get overloaded.

## 6. Build the circuit board.

Transpose your circuit to a perf board and do point-to-point wiring (**Figure L**). The largest round PCB from the 5-pack specified in the Materials list fits perfectly in the lamp base.

## 7. Add power cord and dimmer.

For the power supply wire, drill a ³⁄₁₆" hole in the back of the base. Cut the plug off the cord, feed it through the hole, and tie a strain-relief knot inside. Use your multimeter to check which wire is positive and which is negative before soldering. Then you can solder it directly to the perf board as shown in Figure L.

Drill a hole to fit your potentiometer's threaded bushing in either the front or the back of the base, depending on where you prefer the dimmer knob to be. Insert the pot, mark where the tab goes, and drill a ⅛" hole for the tab (**Figure M**).

Test your lamp, then mount the circuit board in the base using a few dabs of epoxy or hot glue (**Figure N**).

## 8. Add a base plate.

You'll need a base plate to keep your lamp from falling over. I opted for a piece of ABS plastic sheet.

If you use plastic, mark where you want the lamp to sit, then use the circle cutter to cut a hole to fit your lamp base (the PVC coupling). Use epoxy to affix the lamp to the base plate (**Figure O**).

## Enhancements

If you sanded the neck of the lamp, it will be slightly rough and will attract dirt. Paint it with polyurethane for a glossy finish.

Your lamp is complete — for the time being, anyway. Six years from now, you may be able to build an even smaller version, replacing the 18 LEDs with just one. Either way, the freaky purple hue that those big old 1cm LEDs emitted back in 2006 will be nothing more than a memory. ◪

---

Charles Platt is the author of *Make: Electronics*, an introductory guide for all ages. A contributing editor of MAKE, he designs and builds medical equipment prototypes in Arizona.

Gregory Hayes (L–O)

# 1.2.3. Label-Etch a Glass Bottle

by **Sean Michael Ragan** Illustrations by **Julie West**

**HERE'S A SIMPLE TRICK I DISCOVERED** for etching designs on glass bottles using the bottle's label as a built-in resist.

## 1. Prepare the bottle.

This process requires a bottle with an adhesive plastic label. A sure sign that the label is suitable is that parts of it are transparent. » Use a permanent marker to draw your design on the label. » Using your hobby knife, carefully cut around the edges of your design. Lift the edges of the cutout areas using the blade, and finish peeling off each positive cut using tweezers. » Wipe the cut stencil with a paper towel generously soaked with rubbing alcohol. This will remove residual ink and clean any remaining adhesive from the cutout areas. » To make sure the remaining stencil is firmly adhered to the bottle everywhere, wrap a scrap of paper around the bottle, over the label, and rub it briskly with the side of your marker.

## 2. Apply etching cream.

Generously daub etching cream over the exposed positive areas of your design using a brush. » Leave the etching cream in place 5 minutes, or whatever the instructions say, and then wash away all traces of the cream with plenty of warm water in the sink.

## 3. Remove the label and clean.

Using your hobby knife or just your fingernail, lift one corner of the label and peel it off. » Give the etched design one final cleaning with rubbing alcohol and a paper towel to remove any leftover adhesive. ◪

➕ See more step-by-step images at makeprojects.com/project/l/179.

Sean Michael Ragan is technical editor of MAKE. He's descended from 5,000 generations of tool-using hominids. Also he went to college and stuff.

**YOU WILL NEED:** Glass bottle with adhesive plastic label » Permanent marker » **Hobby knife** » Tweezers, small » **Scrap paper** » Paper towels » **Rubbing alcohol** » Etching cream » **Paintbrush** » Safety goggles » **Gloves** » Sink

**WARNING:** Glass etchants are toxic and should be handled with care. Wear gloves and goggles and follow the label directions closely.

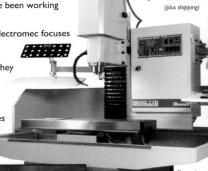

# DANGER Burn Things with a Magnifying Glass

Written and illustrated by *Gever Tulley* with *Julie Spiegler*

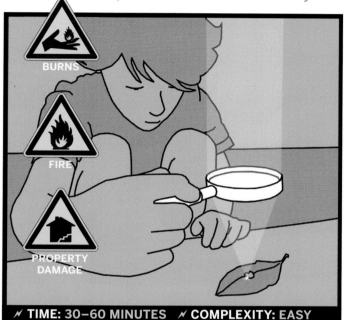

BURNS

FIRE

PROPERTY DAMAGE

⚡ **TIME: 30–60 MINUTES**   ⚡ **COMPLEXITY: EASY**

## Harness the awesome power of the sun.

1. **Prepare.** Find or make an area that's free of flammable materials. A sidewalk, driveway, or dirt path is ideal.

2. **Focus the light.** Hold the lens above the paper and notice the bright circle of light that it makes. Move the lens up and down until you make the smallest circle of light possible – this is concentrated sunlight and it's very hot.

3. **Burn.** Hold the lens still and observe the effect on paper. Try the same procedure on fresh fruit. Write your name on a wood scrap or melt a bit of plastic. Experiment.

The amount of heat you generate at the focal point of the lens depends on the size of the lens and the angle of the sun in the sky. The atmosphere absorbs and reflects some of the light. When the sun is low on the horizon, the light must travel through more of the atmosphere to reach you. ⚡

Excerpted from *Fifty Dangerous Things (You Should Let Your Children Do)* by Gever Tulley with Julie Spiegler (fiftydangerousthings.com). Gever is co-founder of Brightworks, a new K–12 school in San Francisco (sfbrightworks.org).

**YOU WILL NEED:** Magnifying glass, available at drugstores » Scratch paper » **Fresh fruit**

## WARNINGS

You are responsible for extinguishing anything that you ignite. Always work in a clear area where you can't accidentally ignite anything else. » Light hot enough to burn paper can also burn you – so don't focus it on anything you don't intend to ignite.

## NOTE

Treat the magnifying glass gently. The lens can be scratched easily, reducing its effectiveness.

## SUPPLEMENTARY DATA

The speed of light is a constant — in a vacuum. Light travels at different speeds through different materials, so when it goes from one to another (from air to glass) it changes direction in a predictable way. This phenomenon is called "refraction" and it's what enables a lens to focus light, or drops of water to make a rainbow.

It seems remarkable that anything is transparent. After all, glass is more dense than wood and yet somehow visible light can go right through it. Air, water, plastic, and certain minerals are about the only substances transparent to visible light. However, everything is transparent to some form of electromagnetic radiation. Our bodies are transparent to X-rays, the planet is transparent to cosmic rays, and paper is transparent to microwaves.

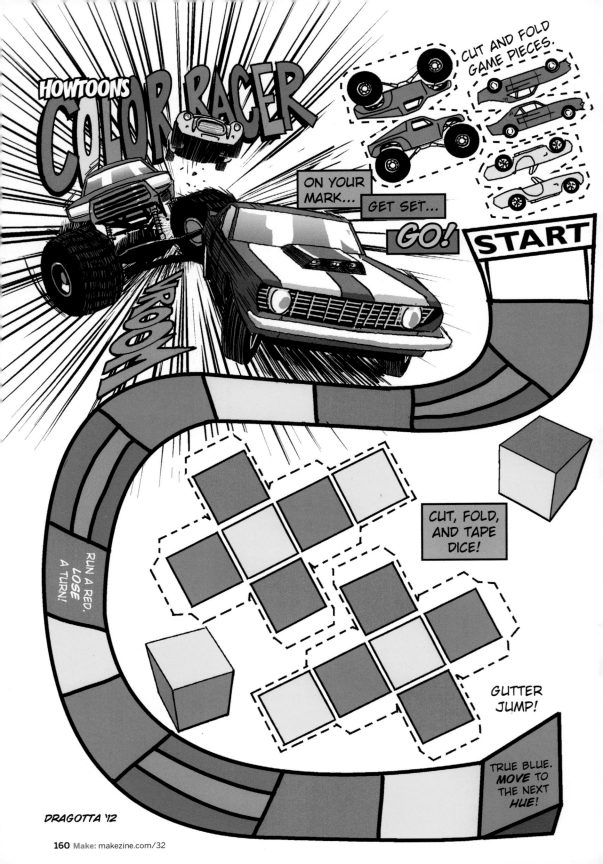

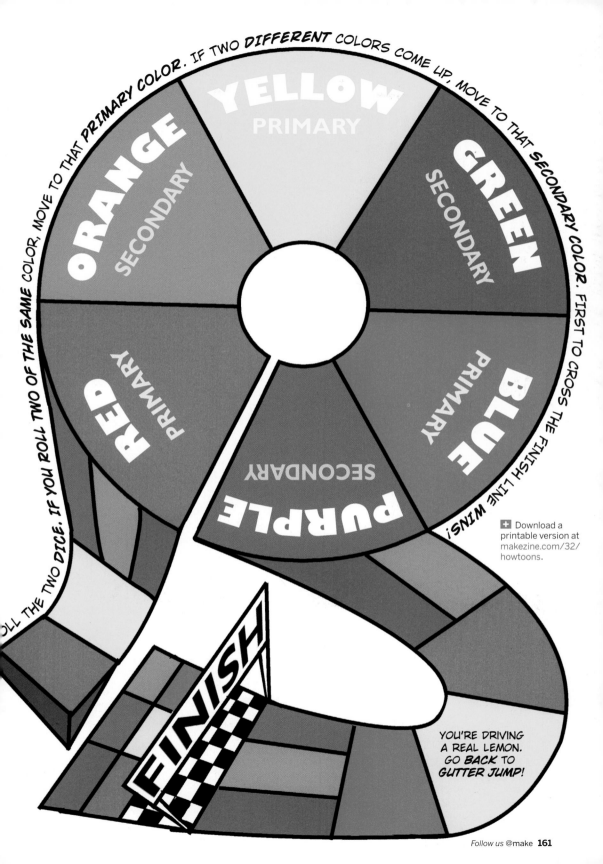

ROLL THE TWO **DICE**. IF YOU ROLL TWO OF THE **SAME** COLOR, MOVE TO THAT **PRIMARY COLOR**. IF TWO **DIFFERENT** COLORS COME UP, MOVE TO THAT **SECONDARY COLOR**. FIRST TO CROSS THE FINISH LINE **WINS!**

**YELLOW**
PRIMARY

**ORANGE**
SECONDARY

**GREEN**
SECONDARY

**RED**
PRIMARY

**BLUE**
PRIMARY

**PURPLE**
SECONDARY

FINISH

YOU'RE DRIVING A REAL LEMON. GO **BACK** TO **GUTTER JUMP!**

Download a printable version at makezine.com/32/howtoons.

# what I made

A CAFFEINE-INSPIRED BEAN-SHAPED COFFEE HOLDER MADE FROM COFFEE-STAINED STIRRING STICKS.

A QUADRUPLE WHAMMY OF COFFEE COOLNESS THAT'S EASY TO MAKE.

TINSNIPS

EPOXY ADHESIVE

6MM

WOODEN BEADS

**1** DRILL 1.5MM HOLES IN BOTH ENDS OF 46 STIRRING STICKS. WIGGLE THE DRILL TO ELONGATE EACH HOLE.

2.6mm

176 mm

**x46** STIRRING STICKS

· 320 MM OF WIRE. SHAPED LIKE A COFFEE BEAN

**2** USE PLIERS TO BEND 1.5MM WIRE INTO A COFFEE BEAN (OVAL) SHAPE.

**3** SLIDE THE STICKS AND BEADS ONTO THE WIRE AND JOIN THE ENDS WITH EPOXY ADHESIVE.

EPOXY GLUE

EPOXY GLUE

STENCIL

COFFEE

COFFEE

EPOXY GLUE

**4** CUT A BASE FROM THE SIDE OF A SODA CAN USING TINSNIPS. ENSURE IT FITS SNUGLY WITHIN THE HOLDER AND SECURE WITH EPOXY ADHESIVE.

**5** NOW FOR THE LABEL. WITH THE AID OF A PAPER STENCIL, TAP HOLES INTO A RECTANGULAR PIECE OF SODA CAN USING A HAMMER & NAIL. ATTACH IT TO THE HOLDER USING MORE GLUE.

**6** FINALLY, MAKE A WOOD STAIN BY BOILING SOME GROUND COFFEE IN...

...A SAUCEPAN. APPLY THE STAIN USING A SMALL PAINTBRUSH.

BY SCOTT BEDFORD
WHATIMADE.COM

WINNER
THE 15TH ANNUAL
webby
AWARDS

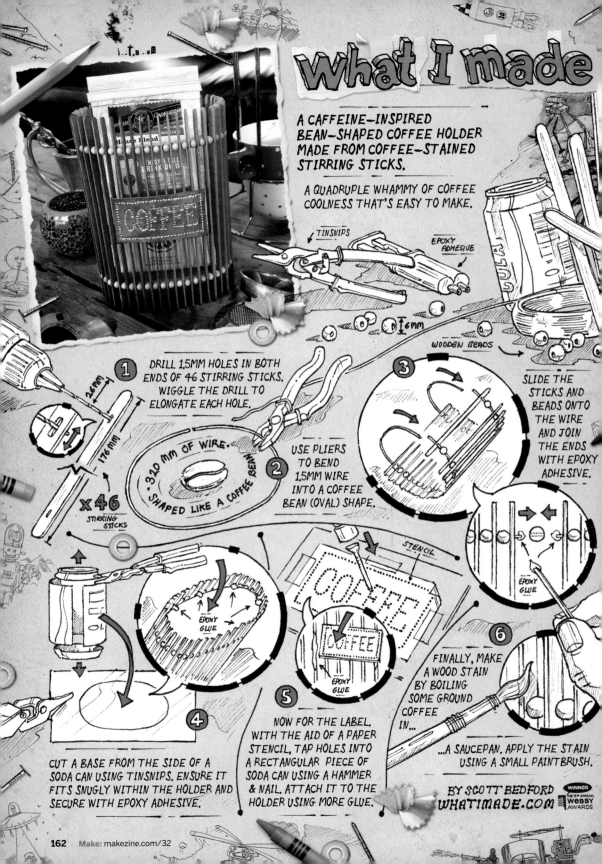

# Make: Marketplace

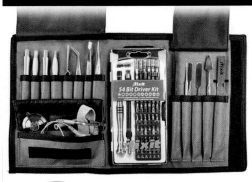

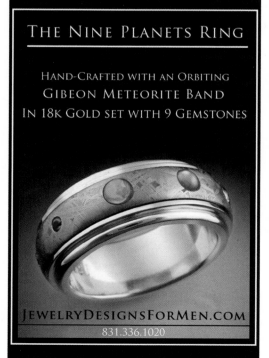
1. Publication Title: Make Magazine; 2. Publication Number: 1556-2336; 3. Filing Date: 10/01/12; 4. Issue Frequency: Quarterly; 5. Number of Issues Published Annually: 4; 6. Annual Subscription Price: $34.95; 7. Complete Mailing Address of Known Office of Publication: O'Reilly Media/MAKE, 1005 Gravenstein Hwy North, Sebastopol CA 95472; 8. Complete Mailing Address of Headquarters: same; 9. Full Names and Complete Mailing Addresses of Publisher, Editor, and Managing Editor: Publisher: Dale Dougherty, Editor: Mark Frauenfelder, Managing Editor: Melissa Morgan, all at O'Reilly Media/MAKE, 1005 Gravenstein Hwy North, Sebastopol CA 95472; 10. Owner: O'Reilly Media, Inc., 1005 Gravenstein Hwy North, Sebastopol CA 95472; 11. Known Bondholders, Mortgagees, and Other Security Holders Owning or Holding 1 Percent or More of Total Amount of Bonds, Mortgages, or Other Securities: Tim O'Reilly, O'Reilly Media, 1005 Gravenstein Hwy North, Sebastopol CA 95472; 12. Tax Status: [x] Has Not Changed During Preceding 12 Months; 13. Publication Title: Make Magazine; 14. Issue Date for Circulation Data Below: July 2012 (Vol 31); 15. Extent and Nature of Circulation, Avg. No. Copies Each Issue During Preceding 12 Months/No. Copies of Single Issue Published Nearest to Filing Date; a. Total Number of Copies (Net Press Run): 123,572/125,831; b. Paid Circulation (By Mail and Outside the Mail) (1) Mailed Outside-County Paid Subscriptions 60,944/58,799, (2) Mailed in-county Paid Subscriptions 0/0, (3) Paid Distribution Outside the Mails 18,468/18,810, (4) Paid Distribution by other Classes of mail through the USPS 0/0. c. Total Paid Distribution (sum of 15 b, (1), (2), (3), and (4)) 79,412/77,609 d. Free or Nominal Rate Distribution (1) Outside-County Copies 414/602, (2) In-County Copies: 0/0, (3) Mailed at other Classes through the USPS: 0/0, (4) Distribution outside the Mail: 2,542/2,821; e. Total Free or Nominal Rate Distribution (Sum of 15d (1), (2), (3), and (4)): 2,955/3,423; f. Total Distribution (Sum of 15c and 15e): 82,367/81,032; g. Copies Not Distributed: 41,205/44,799; h. Total (sum of 15f and g): 123,572/125,831; j. Percent Paid (15c divided by 15f): 96.41%/95.78%; 16 Publication of Statement of Ownership: [x] Publication Required. Will be printed in the March issue of this publication. 17. Signature and Title of Editor, Publisher, Business Manager, or Owner [signed] Heather Cochran, Business Manager, 10/01/12. I certify that all information furnished on this form is true and complete. I understand that anyone who furnishes false or misleading information on this form or who omits material or information requested on the form may be subject to criminal sanctions (including fines and imprisonment) and/or civil sanctions (including civil penalties).

# TOOLBOX

## Schröder Hand Drills $26–$48 garrettwade.com

» When I was a boy, my grandfather was the handyman in the family. I followed him around with fierce devotion and helped him build, fix, and finish things. We were in no hurry, and we liked it. That was a very long time ago. But when I first picked up the Schröder, it felt like shaking hands with a long-lost old friend. If I'd been blindfolded, I might have sworn it was Grandpa's well-maintained vintage tool. The solid construction and tight engineering hark back to the days when tools weren't considered disposable.

*–Gregory Hayes*

© Gregory Hayes

# AmeriKit Learn to Solder Kit

*$19* **makershed.com #MKEL4**

AmeriKit includes everything you need: wire cutters, solder, electrical components, and of course, the soldering iron itself. The kit's instructions are very detailed, and the practice circuit is fun to build. I have built and fixed several things since I got the kit. Now, even my teachers ask me to fix things around the classroom!

—*Robert M. Zigmund, age 14*

# Die and Grommet Tool

*$15* **sailrite.com**

For making trampolines, awnings, bags, sails, and even shower curtains, these are awesome little tools. You'll need to buy one kit (setting tool and hole-cutting tool) for each size of grommet you plan on using. Small, medium, and large are usually enough for everything.

—*Saul Griffith*

# DeWalt ToughSystem Tool Box

*$55 and up* **dewalt.com**

DeWalt's ToughSystem cases are near-indestructible toolboxes that cradle your tools and equipment against rough handling and environmental conditions. Made from 4mm structural foam, these cases don't bend, crack, or deform like many cheap plastic boxes do when fully loaded.

As a DIYer or maker, you might not need a professional-grade tool case, but they're definitely good to have. These water-sealed cases might also serve well as outdoor project boxes, and they cost substantially less than Pelican cases.

—*Stuart Deutsch*

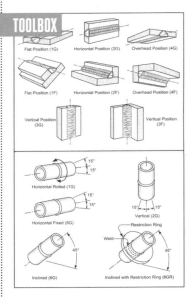

# Welding Know-How
## by Frank Marlow

*$50 **Metal Arts Press***

As a custom luthier I'm frequently called upon to fabricate metal parts for nonstandard guitars and other stringed instruments. I'm primarily a woodworker, but my sputter box and brazing torch don't gather a lot of dust. Although my metalwork has always been adequate, it never reached the high level of workmanship that I really wanted. Then I found this book. The beautiful line drawings clarify the simple, well-written text.

The amount of shop experience packed into this volume is amazing. There are step-by-step instructions for building jigs and fixtures, and some surprising ways to modify tools so that they actually work. This hefty volume is a keeper, and in the years to come I'll probably wear it out.

—*Ervin Tibbs*

# Adjustable Sail-Making Palm

*$22 **sailrite.com***

A fabulous tool in the category of sewing things is a good adjustable palm. These are necessary to get large needles through heavy fabrics. For small repair jobs, this is all you need — forget the sewing machine. Think of it as a supersized, forceful thimble. Make sure to get some heavy-duty hand needles, both straight and curved, so that you're ready for any job.

—*SG*

## USB MICROSCOPE

*$80 adafruit.com*

Visual inspection of surface-mount electronics can be challenging. A magnifying glass is helpful, but if you do a lot of microcircuit work or wish to memorialize your project with photos, this approach can seem as old as Sherlock Holmes. Adding a USB microscope from Adafruit will bring your workspace into the 21st century, while making SMD work easier and preventing eyestrain. After using it only a few times you'll wonder what you'd do without it!

–*L. Abraham Smith, N3BAH*

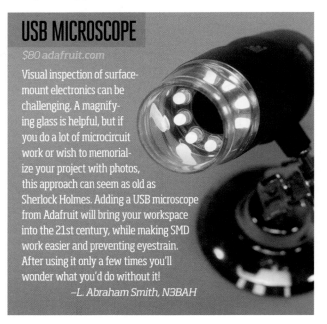

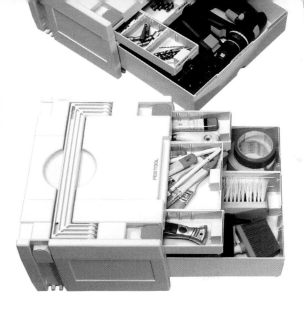

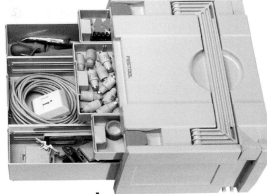

# Festool Sortainer

$147 and up **festoolusa.com**

When I built my first circuit, all of my tools and supplies fit inside a small shoe box. When I built my first robot, all was contained in a small under-the-bed clothing tote. Now, more than 12 years and many projects, tools, and parts later, my storage needs have become so much more complex.

In recent years I've tried almost every type of industrial storage product out there — trays, organizers, drawers, cabinets, bins, and totes. In my latest attempt to organize and tidy up my shop space, I moved a number of items into new Festool Sortainers.

In just a couple of days, they've proved to be among the most versatile and effective small tool and parts organizers I've ever used. They're stackable, portable via a large carry handle, and feature dividable drawers that latch closed and open quickly via finger pulls.

It took me a few tries to arrange and separate the drawers in a way that works best for the intended contents, but that's always part of my new-storage-setup process. The hardest part was working with a limited number of Sortainers while wanting a couple more to fill up.

Sortainers are a bit pricey at over $140 per unit, but their quality, utility, and elegance justify the one-time investment. Although designed for woodworkers, contractors, and professionals, Sortainers are well suited for the more general needs of DIYers and makers. While they won't magically improve your abilities or craftsmanship, they will help organize your shop or workspace efficiently and with flair. —SD

# CULINARY BLOWTORCH

**$26** makezine.com/go/torch

I have a new favorite tool. I needed to make a few solder joints on a ham radio antenna in my backyard. My daughter's culinary blowtorch, purchased for making crème brûlée, was the perfect solution: lightweight, pencil flame, easy to control. I have since seen the same basic torch packaged and sold as a workshop soldering tool at the local hardware store, but this one solders antennas and also makes great crème brûlée.

—*Eric Hansen*

Gregory Hayes

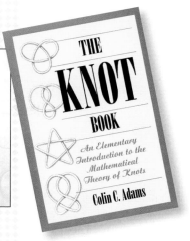

## Design for Hackers by David Kadavy

*$40 **Wiley***

I love design books that attempt to teach the overarching principles of design literacy. Everyone should understand these fundamentals and how the visually designed world is put together. In this vein, David Kadavy decided to address these issues specifically to hackers, because, as he says, they're the ones who are currently reinventing the world.

Translating these concepts into the language and concerns of makers and hackers, *Design for Hackers* looks at color theory, proportion and geometry, size and scale, white space, typographical principles, and overall design and composition principles. Utilizing a technique common in the tech world, Kadavy uses "reverse engineering" to deconstruct and examine real examples of these principles. I think the world would be a far more attractive place if those creating our interactions with it knew the language of design. And, given the extent to which our world has become digital and virtual, those coding its software and user interfaces and threading the web should all learn what this book has to teach.    —*Gareth Branwyn*

## The Knot Book
by Colin C. Adams

*$17 **Henry Holt***

While the systematic study of knots has been passed back and forth between scientific and mathematical disciplines for over a century, most of the discoveries have actually happened in the last 15 years. Colin C. Adams earnestly reveals the accessibility and advanced potential of this field of math. Whether it's learning knot theory that has contributed to our understanding of DNA, or twisting your brain around in new ways, Adams makes you feel like part of the discovery.

—*Meara O'Reilly*

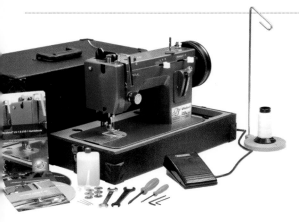

## Ultrafeed LSZ-1 Industrial Sewing Machine

*$900 **sailrite.com***

I happen to think sewing is one of the most versatile and important of manufacturing techniques. I particularly love a strong, industrial, walking foot machine that can handle thick, heavy, and demanding fabrics, even leather. This one even has a nice hand-cranking wheel option that allows you to do heavy fabric repairs off the grid!    —*SG*

## SENKICHI GOLD BUTTERFLY SCISSORS

*$53* **makezine.com/go/scissors**

I have owned several disappointingly cheap pairs of classic Chinese "butterfly scissors" in low-grade carbon steel, and they tend to rust or wear out pretty quickly. This is my first pair in stainless steel, and they are a complete pleasure, both to look at and to operate. I received mine as a gift, and though I don't have to use them very often, I consider them one of the best gifts I've ever been given. —*Sean Ragan*

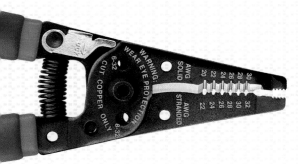

## KLEIN-KURVE WIRE STRIPPERS

*$18* **kleintools.com**

Even high-quality wire strippers can sometimes be a bit finicky when stripping the small-gauge wires used in most hobbyist electronics projects. Klein's 11057 Kurve strippers are extremely comfortable to use and stop even the smallest wires with ease. If you use cheap multi-function strippers and you struggle to cleanly strip small wires, this is the model to upgrade to. —*SD*

## WILDE SLIP JOINT PLIERS

*$15* **wildetool.com**

Wilde recently redesigned their slip joint pliers with a field-serviceable flush-fastener pivot that slims down the pliers for improved tight-quarters access. Some might call slip joint pliers obsolete or old fashioned, and they might be right. Even so, these ubiquitous pliers are undeniably versatile. Sure they seem a bit Spartan, but they're very well made, durable, comfortable to use, and they're made in the United States. —*SD*

## New from MAKE and O'Reilly

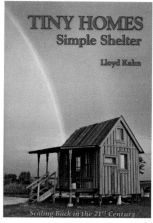

### Getting Started with MakerBot
by Bre Pettis and MakerBlock

*$15 O'Reilly Media*

Get a hands-on introduction to the world of personal fabrication with the MakerBot, the most popular rapid prototyper. Learn how this open source 3D printer democratizes manufacturing and brings the power of large factories right to your desktop. You'll also get guidelines on how to design and print your own prototypes.

### Make: Lego and Arduino Projects
by John Baichtal, Matthew Beckler, and Adam Wolf

*$35 O'Reilly Media*

Make amazing robots and gadgets by combining two of the hottest DIY technologies: the venerable Lego and the upstart Arduino. You'll learn how to take Lego Mindstorms components — motors, sensors, and more — and interface them with the Arduino microcontroller, opening many exciting new options.

### Tiny Homes
by Lloyd Kahn

*$14 (ebook) Shelter Publications*

While my partner and I built our tiny house in the country, we kept a close watch on Lloyd Kahn's blog, where he posted prolifically about making his gorgeous new book, *Tiny Homes*. Kahn has long documented the gems of the DIY house-building movement, but the houses in this book are even more inventive and unique to make up for their lack of space. —*MOR*

## TRICKS OF THE TRADE  By *Tim Lillis*

Here's a great trick to keep small items safe from the elements. Thanks to Brian Green, who originally shared the trick on his backpacking blog at briangreen.net.

Remove the tops of two identical plastic bottles just below the rim near the cap. Sand or file down the excess plastic so that the rim is a clean flange with a flat edge.

Attach the two bottle tops to each other by placing the flattened flanges together, running a small bead of cyanoacrylate glue in between the parts. Let dry in a clamp or vise.

Place your valuables inside, and screw on the two original soda caps for a waterproof container. To create two compartments, add a thin plastic layer between the two flanges.

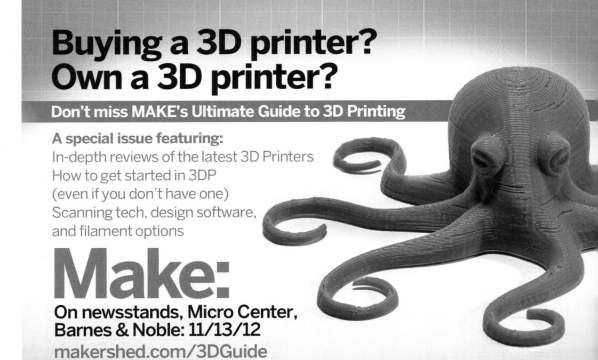

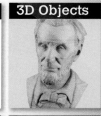

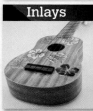

Invented & drawn by *Bob Knetzger*

# TOY INVENTOR'S NOTEBOOK

## Hack in a Hat

Often when pitching a company an idea for a new product, it's not what you have, it's what they think you have. That was the case when I presented this "game-in-a-hat" toy idea. My idea: a preschooler-sized hat that uses a motion sensor and sound circuit to play a simple musical game. As the hat plays a song, the kid dances along (and the motion sensor monitors the child's actions). When the music stops, the dancer must freeze. If he moves – RAZZZZZ! – he's called out. Simple and fun!

But how to make a compelling presentation without having to build a working prototype? Instead, I made a hat with a speaker and pushbutton switch, both wired to an iPod through a wired control (such as the Apple earbuds with remote: makezine. com/go/earbuds). My switch was twinned to the circuit board traces for the "next track" button on the remote, and the speaker was connected to the audio out. Every time I pressed the button on the hat, the iPod would skip ahead to the next audio track!

I also recorded multiple audio tracks (including some silent "spacer" tracks) and made a cleverly sequenced playlist that simulated how the real toy would play the games. The voices and music played through the hat's speaker while I'd surreptitiously press the button to trigger the next audio track as needed to demonstrate the game. As long as I didn't deviate from my canned routine, the musical hat demoed exactly like the real thing. I wore the hat and danced or froze through my demo. The toy company loved it! ◪

➕ To see the TV commercial for the final toy, go to makezine.com/go/dancehat. To see pictures of the prototype go to makeprojects.com/v/32. What cool hacks can you create from an iPod?